The Global Environmental Crisis

*A global perspective for a sustainable future and
a non-exclusive proposal to the European Union*

Giovanni Rantucci

Copyright © 2012 by Giovanni Rantucci
Al rights reserved
ISBN-10: 1439268649
EAN-13: 9781439268643

**Front Page Cover Picture: Earth's
Photo from EUMETSAT, METEOSAT
4 September 2001.**

πάντα ῥεῖ

"Panta rei" (everything flows)

Only change is real, *"like the continuous flow of water, thus you cannot step into the same river twice"*.
Heraclitus (545–475 BC) of Hephesus (Magna Grecia, coast of Asia Minor), is the pre-socratic Greek philosopher famous for his saying, that change is central to Universe. His short sentence struck my imagination as a high-school student and became more impressive when I decided to study geology and confront the restless interplay of the Earth, biosphere and life within the solar system.

Inspiration

The book is dedicated

to the American Julia Butterfly Hill (Luna) for her civil disobedience; Julia spent 738 days (December 10, 1997 to December 18, 1999) in a California Redwood tree (about 1,500 years old) to prevent loggers from cutting it down, and

to Roberto Saviano, the Italian author of *"Gomorrah"*, the book which describes the illegal activities of the Camorra and depicts a community in which people and the environment fall prey to organized crime, while government authorities are absent.

Julia and Roberto strived against injustice through civil disobedience and their personal commitment to preserving the natural environment and human dignity in the absence of which life makes no sense.

Acknowledgments

I wish to express my deepest thanks to colleagues and friends who kindly accepted to read, criticise and revise the manuscript of this book. As a result of the different topics dealt with, the revision was carried out on the basis of a homogenous assemblage of chapters, depending on the background of the reviewers.

This demanding job was carried out by Robert Brinkman, Danilo Severini, Giorgio Cancelliere, Marco Baglioni, Vittorio Panei, Rodolfo Zoppis, and Carlo Battista who provided suggestions on scientific aspects. Giovanna Montagna made the preliminary revision of some chapters, mainly concentrating on the correctness of the language, Valeria Marinucci provided useful suggestions on editing. Fiona Attwood and Alessandra Rantucci made the final revision while my son Lorenzo, helped me with practical advice and formatting. Finally, I would like to thank my wife Roberta for patiently supporting this project.

A number of authors kindly authorized the use of their figures on various subjects. Instrumental was the availability of information from Wikipedia and other sources. The views and findings of some international organizations, whose policies are central to the global environmental debate, are described and supported by figures essential to the structure of the book:

- IPCC (UNEP)
- WWF
- EREC/Greenpeace
- IEA/WEO
- MILLENNIUM ECOSYSTEM ASSESSMENT

My special thanks goes to Herman Daly, American Ecological Economist, Professor at the School of Public Policy, University of Maryland, College Park USA, who kindly authorized the reproduction of figures from his book "Beyond Growth" (1996). His criticism of traditional economics and "business as usual", fostered my view that the global environmental crisis is mainly a problem of social development, knowledge, awareness and moral values.

Contents

Inspiration

Preface

CHAPTER 1 – THE UNPRECEDENTED THREAT OF THE GLOBAL ENVIRONMENTAL CRISIS	1
The Global Environmental Crisis	1
The Most Critical Issues	3
The Decline of Ecosystem Services	5
Climate Change	7
Historical Development of the Crisis	9
The Transition of the Past 40 Years	10
The Structure of the Book	12
CHAPTER 2 – COSMOLOGY	15
Our Cosmic and Terrestrial Environment	15
The Big Bang	17
Cosmological Constants and the Anthropic Principle	20
The Formation of Chemical Elements	21
Mendeleev table, origin of chemical elements and the evolution of stars	*21*
The origin of carbon	*24*
CHAPTER 3 – THE EARTH AND THE BIOSPHERE	27
Solar System, Planets, Asteroids and Comets	27
The Earth's Geological and Biological Evolution	31
The Earth's Magnetic Field	37
The Biosphere	38
Size, composition, environment, ecosystem and living organisms	*38*
The tree of life	*39*
Requirements for the Formation of a Biosphere and the Rise of Intelligence	41

CHAPTER 4 – FROM PREHOMINIDS TO HOMO SAPIENS   45
  Prehominids and Hominids   45
  The Wurm Ice age, Farming and Industrial Revolutions   48
  The Migration of Homo Sapiens   50

CHAPTER 5 – GLOBAL WARMING AND CLIMATE CHANGE   53
  Climate Change Factors   53
    Introduction   53
    Extraterrestrial factors   54
    Terrestrial factors   57
    Human factors   58
  Research Methods and Effects of the Climate Change   58
  Climate Change   60
    The cenozoic era   60
    The paleocene-eocene thermal maximum or PETM   64
    The 2°C overshooting and temperature inversion during ice ages   66
    The last ice age   69
  The Atmosphere's Energy Balance and Global Radiative Forcing   72
    Energy flow and the greenhouse effect   72
    Anthropogenic energy contribution and the global radiative forcing   76
    Temperature variation, ice melting, the thawing
      of tundra and permafrost   78
  The Thermohaline Circulation Shutdown   79
  Conclusive Remarks   81

CHAPTER 6 – A HISTORICAL VIEW OF CAPITALISM
AND DEMOCRACY   83
  The Evolution of Capitalism   83
  Democracy as an Unfinished Process   85
  Historical Views by Smith, Malthus, Mill and Clausius   86
  Criticism of Lobbies and the Traditional Market Economy   90
  Emerging Asian Powers and Global Effects   92
  Conclusive Remarks on Capitalism and Democracy   93

CHAPTER 7 – ECONOMIC GROWTH
AND SUSTAINABILITY   95
  The 1987 Brundtland Report and United Nations Involvement   95
    Introduction   95
    The Brundtland Report   97
  The Sustainable Development by Herman Daly   99
    A voice in the darkness   99

    Orthodoxy and ignorance   *100*
    Concepts and definitions   *101*
    The importance of the carrying capacity and the Plimsoll Line   *103*
    Biophysical, ethical and social limits   *104*
    Circular flow, economic and uneconomic growth   *104*
    Gross national product and externalities   *108*
    Economic growth, globalization and daly's recommendations   *109*
  Conclusive Remarks   111

CHAPTER 8 – SCENARIOS FOR THE 21ST CENTURY   113
  Global Scenarios   113
  The Millennium Ecosystem Assessment Scenarios   114
  The Scenarios Associated with the Initiative by the Club of Rome   116
    The trilogy and key concepts   *116*
    Scenarios published between 1972 and 2004   *118*
    Concluding remarks   *120*
  The Ecological Footprint Analysis   121
    Human footprint, bio-capacity, overshooting and ecological debt   *121*
    Ecological footprints, bio-capacities and freshwater consumption   *124*
    Implications of current human development and scenarios   *128*
    Pros and cons of the ecological footprint analysis (efa)   *130*
  The Ecological Economics Approach   131

CHAPTER 9 – THE SUSTAINABILITY DEVELOPMENT PROCESS AND THE REFERENCE TRANSITION MODEL   133
  Humanity in a Phase of Transition   133
  The Nine Major Indicators of the Crisis   135
  The State of the World   137
  Sustainable Development and the United Nations   140
  The Involvement of Nations in the Reference Transition Model   142

CHAPTER 10 – THE ENERGY REVOLUTION   145
  Energy Sources and Carbon Dioxide Emissions   145
  The 2007 Evaluation of Different Scenarios   147
  The Adv Energy (R)evolution Scenario 2010 by EREC/Greenpeace   150
    Assumptions, key principles and new policies   *150*
    The advanced energy (r)evolution versus the reference scenario   *152*
    Some final considerations on the adv e/r scenario 2010   *156*
  The World Energy Outlook 2010   158

    Tackling climate change and securing a sustainable future   158
    The puzzle of components and the
    IEA-WEO 2010 major scenarios   *159*
    The oil peak and coal-fired electricity generation   *162*
  Concluding Remarks   165

CHAPTER 11 – THE SUSTAINABLE DEVELOPMENT OF INFRASTRUCTURES   169
  The Pathway Towards a Sustainable Society   169
  Infrastructures for a Sustainable Development   170

CHAPTER 12 – THE CULTURAL REVOLUTION   173
  Wisdom, Ignorance, Scepticism and Misinformation   173
  The Background of the Global Environmental and Financial Crises   175
    The complex history of the last 100 years   *175*
    The difficulty of changing the present and figuring out the future   *176*
  The Cultural Revolution   178
  The Waning of Humaneness   181

CHAPTER 13 – A GLOBAL PERSPECTIVE FOR A UNILATERAL ACTION OF NATIONS AND A PROPOSAL TO THE EUROPEAN UNION   183
  Introduction   183
  The RTM Proposal to the EU Compared with
    other Regional-size Nations   185
  The Global and Regional Problems of the EU   186
  The Proposal to the EU   187
  Environmental Ethics and the Need for a Rapid Transition   189
  Knowledge, Awareness, Indignation and Participation   191
  Conclusion   193

  References   197

  Links   201

  Index   205

  Glossary   209

  Acronyms and Abbreviations   217

  Units of Measurements   219

Preface

The global environmental crisis is by far the most debated international problem since it involves a variety of conditions affecting humanity in its most critical transition stage. The book is the result of an effort to focus on environmental problems, energy, economy and the possible way out of the crisis, within the given constraints imposed by population growth, human activities and natural disasters.

The target is to clarify the basic points of the crisis, which is primarily human by involving the understanding of current trends in the socio-economic domain, the organization of the social system, the ecological limits and the awareness of the complexity at the global scale. At a lesser extent the crisis is a scientific problem centred on the identification of technological solutions and priorities.

Seven billion people live today on Earth! We have developed into a massive population whose survival generates giant impacts, accelerates the pace of climate change and dramatically affects the progress we have achieved. The current environmental degradation is not the revenge of nature, rather a knowledge and awareness limit that humanity may not be able to overcome in time. Within the last two centuries a bursting scientific progress and new technologies fostered an unprecedented economic growth, based on the exponential rise of population, the consumption of non renewables resources and the increase of impacts. Our society grew indifferent until 1960s to the effects of its fast development, in the last five decades only realizing the dramatic decline of environment and life.

The possibility of a sustainable development process, which is at the moment our only chance for a common survival, is heavily contrasted by the current economic trend, which is in turn supported by the huge consumption of energy and matter and the assumption of an unstoppable economic growth. The complex and unavoidable transition to a sustainable society appears, therefore, a long term process, full of uncertainties and resisted by social and political institutions rooted on the denial of natural limits and the idolatry of an ever rising GDP.

The purpose of this book is to help people open their eyes, understand problems, become aware of the complexity of man and nature interactions and initiate a process of change towards an ecological, democracy-based, sustainable society.

CHAPTER 1

[The Unprecedented Threat of The Global Environmental Crisis]

The Global Environmental Crisis

The beginning of the third millennium calls for an in-depth reflexion on humanity's explosion as a new geological force, at the moment unable to coexist with its own supporting environment. Anthropogenic impacts are seriously reducing the advantages of man's overwhelming progress, causing an unprecedented global environmental crisis, in turn a potential driver of a mass extinction[1] 65 million years after the demise of dinosaurs. Despite the evident signs of crisis, humanity is still blindly heading towards an improbable limitless

---

1   The past five big mass extinctions occurred 434, 360, 251, 205 and 65 million years ago.

survival, which could soon be hampered by the interaction between food scarcity, population growth and climate change[2].

According to a study[3] commissioned by the American Museum of Natural History of New York, if the current decline in biodiversity is not halted, 50% of species will be wiped out within this century. Documented geological and paleontological data show that extinctions of medium and local dimensions repeatedly occurred in the past, associated with climate change or even with abrupt decade-lasting variations in the weather.

The interplay between the ecosystem's evolution and climate change on the other hand is the most important long-term phenomenon in the biosphere. The geological sequence of rock formations and fossils – despite interruptions due to surface erosion and other processes – shows that climate has changed unceasingly in the past and that animal and vegetal life either adapted to changes or disappeared. The imprint of climate and the fossilized remains of life are sculptured in rock sequences worldwide, like words in a natural history book. Sedimentary rocks in particular hold accurate information on the evolution of life and environmental conditions at the time in which fossilized organisms lived. A most remarkable difference, however, qualifies the current global crisis compared to the past five mass extinctions. The major driver this time is humanity which – through innumerable activities and the related impacts – threatens the whole ecosystem. We live in a very "special cosmic laboratory" of finite dimensions, in which the rising complexity of our society and the trend of natural phenomena represent an explosive mix, understandably a source of a harsh debate between pessimists and optimists on the fate of humanity. According to the Stockholm Resilience Centre[4], nine environmental phenomena are close to or beyond threshold limits, drastically lowering the capacity of the Earth to support the sustainability of life and human civilization. For each of these phenomena, pre-industrial values, current values and boundary limits have been identified. The study warns that (i) climate change, biodiversity, nitrogen and phosphorous emission flows into the biosphere have already trespassed limits, (ii) land and freshwater pollution, stratospheric ozone depletion and ocean acidification are beyond the irreversibility limit and, (iii) aerosol loading and chemical pollution are still being evaluated for identification and comparison of values.

---

2   The nearing to 2°C above the mean preindustrial temperature is considered today a most compelling limit.

3   The website www.massextinction.net is devoted to a study (April 1998) commissioned by the American Museum of Natural History of New York, 1998. Based on a survey of 400 scientists, it was updated until 2010 with new articles (and 300 links) containing a great variety of data on the ongoing mass extinction.

4   www.stockholmresilience.org/planetary-boundaries. Stockholm Resilience Centre.

A number of scientists fear that the present crisis may reach its most critical phase between 2015 and 2040, therefore definitely representing an exceptionally short but powerful event. External observers from another planetary system might conclude that the current emergency on Earth is likely to end with another big demise of species. Unfortunately, we are the special observers from within the biosphere, both drivers and victims of our development. Our only chance of survival is to understand the current emergency, identify priorities and take action on the assumption that there is still a possibility to change the current trend. In the last few decades, whenever adopted, environmental protection measures[5] have reduced impacts on the Earth.

The Most Critical Issues

Recent anthropogenic impacts threaten the biosphere as a whole, causing large-scale problems, severe disasters and the decline of human society. The awareness of threats started to grow since 1950s when the intensive use of biocides in agriculture became widespread in America, their negative effects heavily affecting living organisms and their habitats.

Major indicators of the ongoing crisis are:

- population size: 1960 3 billion, 2000 6 billion and 8 billion expected in 2025. In the case of drastic natural changes, adaptation becomes more difficult in a progressively polluted environment, while migration is practically unfeasible in an ever-crowded world

- climate change: greenhouse gaseous emissions, global warming, the accelerated melting of the ice sheets and rising sea levels

- environment and ecosystem degradation: loss of biodiversity[6], habitat destruction, pollination decline, the spread of invasive species and coral

---

5 The drastic reduction in the use of CFC halted the depletion of the ozone layer in Antarctica, while the creation of natural parks and marine reserve areas locally hampered the decline of biodiversity.
6 During the last 500 years, 875 species have become extinct according to the International Union for Conservation of Nature, the extinction of a number of mammals is ongoing, and tens of thousands of undocumented species are estimated to have disappeared during the 20th Century.

bleaching. Driving factors are pollution, farming, cattle grazing and breeding, monoculture and large-scale meat production

- land degradation: desertification, soil pollution, salinization and erosion;

- pollution of water: thermal overheating, ocean/sea dumping and the associated pollution, acid rain, oil spills, ocean acidification, eutrophication of freshwater bodies by untreated wastewater

- air pollution and ozone depletion: GHG and CFC emissions

- resource depletion: dominant use of non-renewable energy sources, slow transition to energy conservation and efficient use

- nuclear energy production: radioactive waste, nuclear accidents, melting and fallout

- impacts on the natural environment: illegal overfishing, deforestation, mining and the associated waste

- overconsumption, overproduction and obsolescence of goods

- politics: inadequate, late and often indifferent to climate change and other related phenomena.

The variety of impacts associated with human progress provides a dark, though not ultimate picture of our future. We are the only species with the ability to modify the environment, through creativity and growing pollution and impacts. During the last few decades the global crisis developed faster than ever and people realized the fragility of the environment. Despite the harsh debate on GHGs emissions and sustainability indicators, the majority of political leaders neither reacted to the critical deterioration of the environment, nor developed a global view for a common survival. The debate, on the type of society in the short term and the global civilization that humanity should build up during this century, is dramatically slow. Above all economic, financial and market issues have risen, during the last

5 years, to the top of the international political agenda, marginally involving global warming, the decline of biodiversity and the increasing pollution. Indicators and their values need to become essential reference points for policy-makers, economists and the civil society, if a globally sustainable system has to be developed for a common survival.

The Decline of Ecosystem Services

Human impacts on the environment became significant since the industrial revolution, growing dramatically faster during the last 50 years. Figure 1.1 shows the interaction between the ecosystem and the human system. Services provided by the ecosystem (supporting, provisioning, regulating and cultural) are essential to life as basic constituents of well-being.
It is worth recalling The Millennium Development Goals (MDG), Progress Report 2011, launched in Geneva by the UN Secretary General (July 2011), which presents the yearly assessment of global human progress and warns that MDGs targets (based on more than 60 indicators) will not be reached by 2015.
The decline of ecosystem services and the delay in reaching MDGs goals dramatically increase human vulnerability. The Earth's ecosystem and humanity represent the most complex "holistic domain", the properties and the dynamics of which cannot be explained by the behaviour of single components or their sum. A holistic system is our body, since its subsystems[7] interact harmonically, being complementary to one another.
The services provided by the ecosystem to life are for free, but our society is reluctant to value them qualitatively, biologically and economically. Cutting down a forest, which hosts a variety of habitats and biodiversity and reduces the erosion caused by running water, it is still too often considered a profitable business (for few)!

---

7   Immune-defence, cardio-circulatory, digestive and other sub-systems are interacting components of our species.

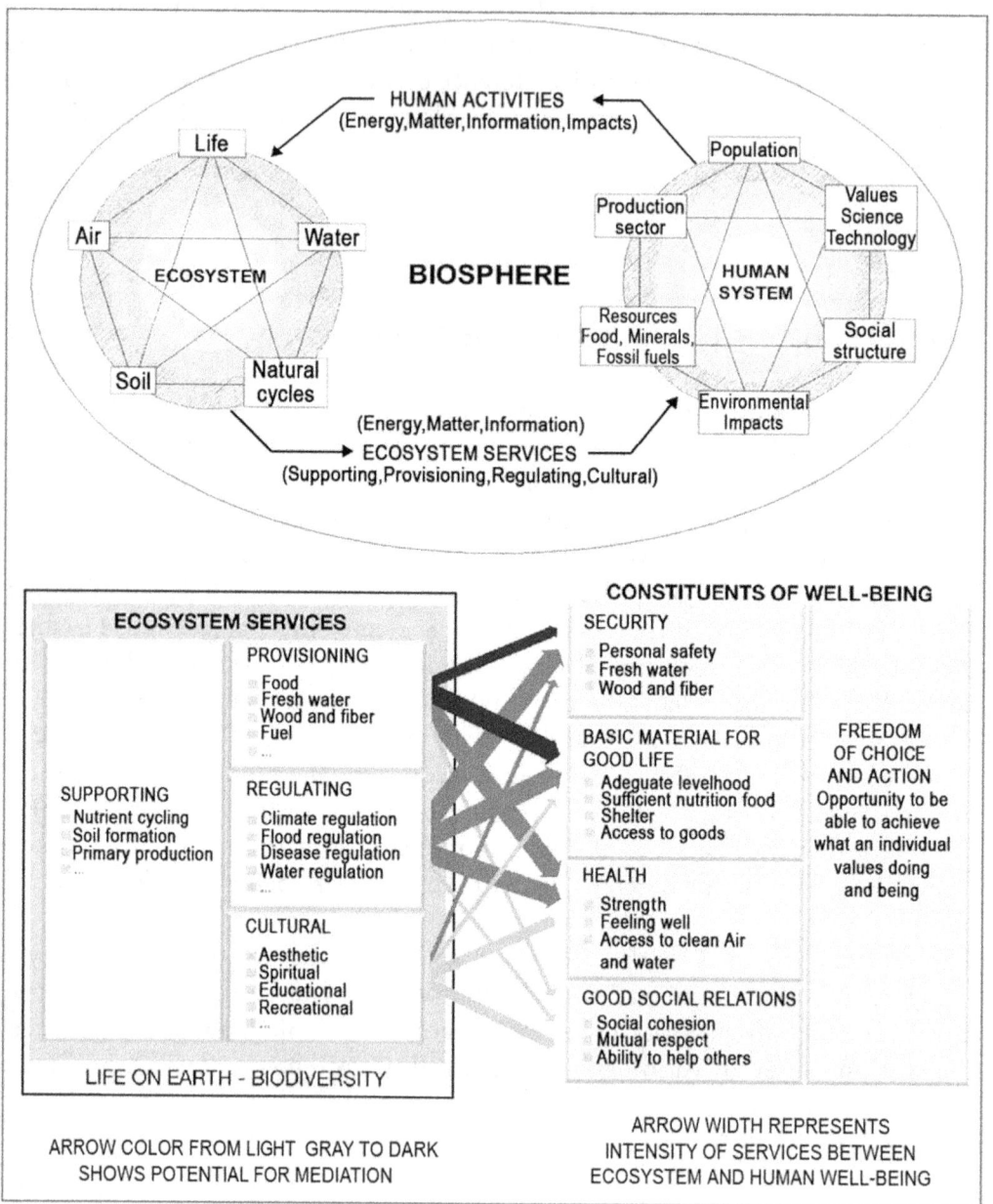

Figure 1.1 – (top) The biosphere and its two interacting domains. (bottom) Services provided by the ecosystem. Source: WRI – Millennium Ecosystem Assessment, 2005.

Climate Change

Climate change is today a central human concern within the environmental domain in which a long-term complex interplay of events, reactions and interactions of components has taken place during the geological history. Traditional drivers of climate change are solar radiation, orbital properties of the planet, tectonic forces, life, natural cycles and human impacts. The variation of the mean global temperature is the major sign of climate change and related effects are

- the melting (or thickening) of ice covers, sea level changes and associated effects on the distribution of living organisms

- the faster modification of the physical world through erosion, weathering of surface rocks, the transportation of sediments by rivers and the accumulation of solid materials in land and water environments

- the evolution of life in the biosphere, which becomes more dynamic at a time of climate change due to the alteration of habitats, the migration of animal and vegetal life, the invasion of species and extinction.

The centrality of climate change and the connection between human activities and the consequences are shown in Figure 1.2. Seen from a short-term human perspective, big disasters appear uncommon events on Earth. In the current case, they are associated with climate-induced anthropogenic alterations, and their frequency and intensity are expected to rise abruptly, with dramatic consequences on environment, ecosystems and human organization. When the preliminary signs of the crisis started being felt in the 1960s, and more significantly at the beginning of the 1970s, climate change was a remote hypothesis. Today, an overwhelming amount of scientific data confirms that we are already in the middle of a climate transition. Greenhouse gas emissions associated with human activities are recognised by scientists to be responsible for global warming, in turn a major cause of climate change. Based on recent data, global surface temperature has increased faster during the past two centuries and it is expected to reach 2°C above preindustrial era within decades.

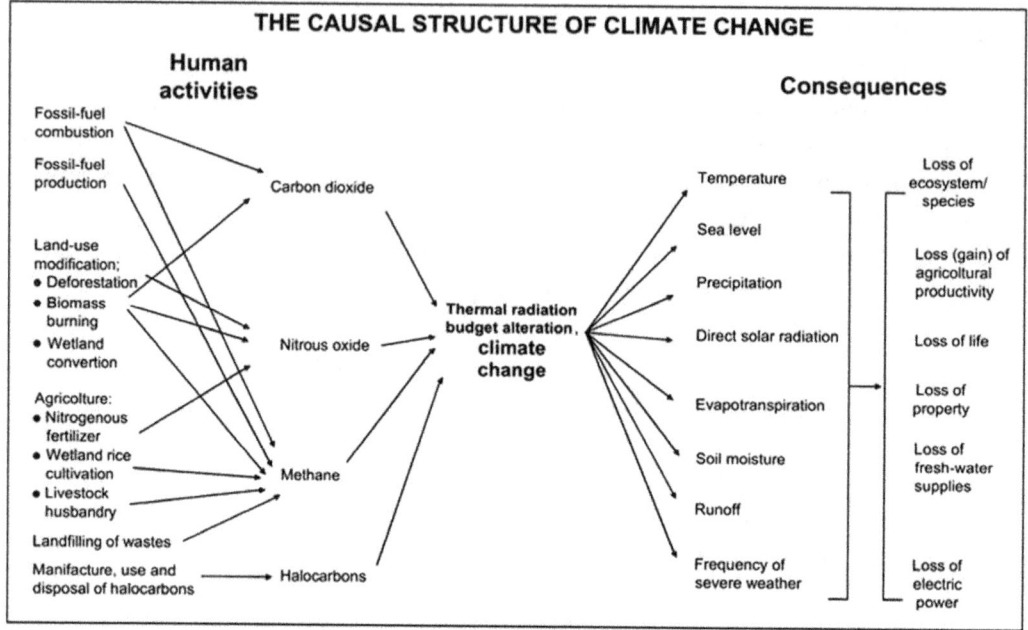

Figure 1.2 – The structure of climate change, adapted from Kasperson and Kasperson, The Global Environmental Risk, 2005. The original source of the figure is used under the permission of the United Nations University Press.

It is very likely that by failing to implement new policies, our society will not be able to cope with the extreme events that climate deterioration will cause.

Lowering gaseous emissions implies a global social change humanity has never faced before:

- the reduction in GHG emissions through a lower consumption of fossil fuels and a higher use of renewables involves adapting industrial processes to a green economy, which in turn requires the modernization of human activities, the protection of the environment and a different organization of the social system

- changes in the lifestyle of people in the direction of conserving rather than wasting energy and matter and lowering food consumption levels in industrialised countries

- changes in government policies and governance.

A widespread knowledge of the biosphere and the environment, and a greater awareness of global issues by the political and economic establishments are indispensable prerequisites for a new trend.

Historical Development of the Crisis

The first acknowledgment of an emerging environmental problem came about in the academic research domain in the late 1950s in America. The growing concern was brought to the public attention by Rachel Carson's 1962 book "Silent Spring", which represented at the national and, later, international level a wake-up call on the use of dangerous and toxic products, thus paving the way for the literature explosion which took place in the early 1970s.

Carson described the environmental pollution due to industrial production, the combined interests of farmers and the chemical industry, and the indifference of the authorities which largely ignored the problem. The main topics in her book concerned the lethal effects on animal life resulting from the use of biocides (pesticides, herbicides and insecticides) in agriculture and the dangerous consequences for humans. The warning by Carson had a shocking effect on American public opinion in the 1960s, and a strong non-scientific opposition attempted to expose her to ridicule.

Articles in her defence were published in the journal Science, and institutions including the Food and Drug Administration (FDA) supported her ideas. The impact of pollution on nature is now a global issue that does not solely concern economically and technologically advanced countries, but in different ways the whole of humanity.

Since the 1970s, scientific research has produced concrete data on the extent of the ongoing environmental degradation, warning that economic growth will be unsustainable in the future. Various government agencies worldwide took the advice seriously and this helped to spread awareness about the environmental problem at an international level.

A variety of books examined environmental, biological, ecological, demographic, economic and political issues including the uncontrolled industrialisation and related impacts.

Philosophers Herbert Marcuse and Bertrand Russell, the economist J.K. Galbraith, and several scientists expressed critical views on the structural organization of modern industrial society. Galbraith took a particularly authoritative position on the free market economy and its effects on the public domain. His criticism concerned the market structure, the rapid development of American industry (which produced goods and services for the welfare of people ignoring the damage to the environment), and the ubiquitous presence lobbies which heavily interfered with politics at the highest decision-making level.

In 1987, the Brundtland Report "Our Common Future" proposed as a moral obligation the transition to a sustainable condition, defined as a development that meets the needs of the present without compromising the ability of future generations to meet their own.

The variety of books and reports published on the environmental crisis during the last decades have traced the evolution of the debate up until now, at the same time representing a labyrinth of studies and data.

In 1972 the Stockholm Summit was held by the UNEP on the Human Environment, and for the first time global, social and environmental issues were discussed at the international level.

The 1992 UN World Conference in Rio de Janeiro on Environment and Development, ended with a convention on GHGs emission control and the approval of Agenda 21; the agreement was signed by most participating countries. The 3rd and 4th Summits were respectively held in 1997 in Kyoto and in 2002 in Johannesburg: great expectations and few results. In a time of emergency, global conferences helped, however, to keep alive the debate on environmental issues.

The UN Millennium Ecosystem Assessment Study — published in 2005 — highlighted the decline of ecosystems, the on going mass extinction and the reduction in the services rendered by ecosystems to society.

The November 2009 Copenhagen Summit and the December 2010 Cancùn Conference, concluding a decade-long debate, recognised climate change as a most dramatic challenge, but did not reach an international binding agreement on the reduction of GHG emissions.

The Transition of the Past 40 years

The decades from 1970 to 2010 represent a fundamental transition period in human history:

- before 1970, the global population reached its maximum growth rate of 2.1% per year, then declining to 1.35% in 2000. Population passed from 3.7 billion in 1970 to 6.9 billion in 2010, it is expected to grow to

8 billion by 2025 and stabilize at 9 billion by 2050. The yearly growth of 74.6 million in 2009 is assumed to decline to 30 million in 2050.

- major global oil crises occurred in 1973, 1979 and 2007, and several oscillations of the market price during the same period, due to conflicts in the Middle East

- between the 1970s and 2010, literature highlighted the effects of global warming, climate change, the decline of ecosystems and governance, which are today among the most serious international problems

- in mid 1970s the ecological footprint[8] exceeded the Earth's biocapacity, this leading to a transition from sustainability to an unsustainable development which in 2011 appears unstoppable

- by the end of 1970s the transition from traditional "capitalism" to "supercapitalism"[9] transformed the free-market into a powerful driver of growing profit, environmental impacts and an increasing threat to civil rights and democracy.

Keeping the Earth hospitable is humanity's major problem now. The change that is needed to halt anthropogenic impacts should be implemented before the system enters a phase of irreversible decline.

Man's sense of ownership of the biosphere and the capacity to modify it were felt as an unquestionable right much before the 1800s when the population was below 2 billion, impacts were limited, the quality of air and water excellent and forests covered the planet. As a consequence of the industrial revolution 200 years ago, rapid scientific progress, huge consumption of energy, the importance of property and the right to use it almost indiscriminately, grew rapidly.

The present condition of humanity and the declining prosperity of the ecosystem depict a planet which is no longer the same, certainly inadequate to support further demographic and economic growth. On the other hand humanity neither can regress to a preindustrial condition, nor continue with the current growing

---

8 The human ecological footprint represents the amount of biologically productive land and sea surface required to meet humanity's demand for products and absorb waste including CO_2 emissions from fossil fuels and other activities.
9 Supercapitalism. The transformation of Business, Democracy and Everyday Life, Robert B. Reich (2007).

trend of the economy. The possibility of deindustrialization is unrealistic, due to the size of the population (7 billion in 2011) and related needs for survival.

The unprecedented development humanity underwent during the last two centuries has fostered the unscientific idea that climate can be considered an almost stable system with peaks and lows, and a pattern similar to seasonal cycles.

Today we should instead (i) accept the scientific finding that climate change is a stable feature of the biosphere's dynamics, (ii) build up awareness that we are subject to the natural challenge of climate, as the other species did in the past, and (iii) be aware that our society has grown too much and too fast in a period of unprecedented improvement of social conditions and declining resources..

The Structure of the Book

The structure of the book is based on the series of consecutive events which lead from the origin of the Universe to the solution of problems related to the Earth's global environmental crisis:

- the Big Bang, the expansion of the Universe, the formation of galaxies, stars, the solar system, Earth and the evolution of biosphere are essential to the understanding of the current crisis, which is rooted both in this background and the rise of Homo Sapiens. The sequence of events since the Big Bang is described in chapters 1 to 4

- the driving factors of climate change, ice ages, the survival of our ancestors during the last ice age, mechanisms of temperature inversion and the recent human-related impacts associated with anthropogenic activities (mainly based on the consumption of fossil fuels), are described in Chapter 5

- three interconnected topics, namely (i) a historical overview of the economy, (ii) the ecologically sustainable development process and, (iii) the reference scenarios for the 21st Century, are respectively illustrated in Chapters 6, 7 and 8

- the current state of the World, the need for a long-term Sustainable Development Process (SDP) to be implemented during this century, and the Reference Transition Model (RTM), as a short-term urgent approach within the SDP to halt climate change and human impacts, are described in Chapter 9

- the three components of the Reference Transition Model, that is the energy (r)evolution, the sustainable development of infrastructures and the cultural revolution, are described in Chapters 10, 11 and 12

- Chapter 13 focuses on a proposal to Nations for a unilateral way out of the crisis and a suggestion the European Union to start first..

The recurring concept of the book is that we live in a complex planetary laboratory in which at the moment life and human survival are in danger. Our species developed intelligence, culture and scientific knowledge, growing in complexity and very recently achieving better living conditions, but also becoming disrespectful of the biosphere's limits and the Earth's ecosystem.

What can we do now? To sum the situation up in a single sentence we could simply say: humanity is in peril, for the first time the risk is global and the solution is in our hands.

CHAPTER 2

Cosmology

Our Cosmic and Terrestrial Environment

The cosmic adventure of *Homo sapiens* began 200 thousand years ago, yet the origins of energy, matter and natural laws date back to the Big Bang, around 13.7 billion years ago. The imaginary journey to the origins of the Universe is the key to understanding the past, unveiling the present and imagining the future.

Concise scientific information on this matter is essential since it helps to understand major steps in cosmic evolution and the formation of the Sun, planets and the Earth in particular, which through astonishing coincidences became suitable for hosting a biosphere. Energy and matter, natural laws and their relationships can be more easily described through the long chain of events that followed the origin of the Universe and its expansion. Life on Earth began soon after the tidy arrangement of the Solar System 4.5 billion years ago, and has evolved unceasingly until today. Six million years ago pre-hominids emerged as the leading mammals, about 200 ky[10] ago *Homo Sapiens* appeared in Africa and his descendants from 100 to 15 ky ago migrated towards other continents conquering the planet. The majority of

---

10   1 ky is equal to 1,000 years

people during past centuries was unaware of living in a planet orbiting around a medium-sized star in the Orion arm of the Milky Way.

As stated by Carl Sagan, in the introduction to the 1988 book "*A Brief History of Time*" by Stephen Hawking, our current life neither takes into account the complexity of mechanisms concerning the energy provided by the sun and the gravity which keeps the components of the solar system into equilibrium, nor the atoms of which our body is made of.

Similarly the problem of "limits" to the future expansion of our fast evolving social system is not considered an integrant part of our basic knowledge. Science in general and Cosmology in particular are considered a prerogative of scientists.

In recent years, however, an alternative knowledge process gained ground: the need for an open-minded approach to questions, such as the origin and evolution of the Earth, the variety of physical and social limits in which we are growingly trapped and the global impact associated with our industrial society.

Dimension and events of the journey from the Big Bang to present were divulgated by Steven Weinberg's famous book "*The first three minutes*" (1977), which paved the way for a different cultural and intellectual approach to the complexity of the universe. The transition of our species from the plough to particle accelerators was accomplished during the last 10,000 years, the understanding of physics and natural sciences being instead mainly concentrated in the last two centuries.

Without looking into the distant past, the current resource basis and the financial, economic, social and environmental aspects of our energy-based society would remain unknown, the present emergency being solely attributed to human misdeeds. We live today in a very unique cosmic environment, that is the biosphere, and in a time-window characterised by high-level scientific knowledge.

Our species is unavoidably involved in a critical transition that could either turn into a global catastrophe or into a major social and cultural step forward in evolution. The urgent need for a third revolution (after the farming and the industrial ones) is now strongly felt, but not shared by the human community due to the different cultural and economic development levels of nations and the blindness of political establishments.

The cosmic environment in which humanity developed is made out of matter and energy, with life originating from a unique combination of six chemical elements: C, H, O, N, P and S[11].

Darwin compared the effort of understanding the evolution of life on Earth through the sequence of fossil remains, to reading a book in which numerous pages

---

11   Carbon, Hydrogen, Oxygen, Nitrogen, Phosphorous and Sulphur are considered the bricks of life.

had been lost and the available pages included incomplete sentences and missing words. At the beginning of the 16th century the understanding of our planet went through isolated discoveries which then led to a correct view of the Solar System and, during the last century, to a holistic, scientifically-based picture of the Earth and the biosphere.

In his book *"Our Cosmic Environment"* (2001), Martin Rees expresses the idea that if extra-terrestrials exist, they are made of the same atoms we are made of, probably oriented towards a similar knowledge process and, in a number of cases, close to discovering the basic trend of cosmic evolution. As he recalls, all living organisms share a common origin: they are made of stardust from nuclear explosions.

The environmental crisis is linked to the cosmic and the recent past in a variety of ways: energy, matter, the forces of nature, the time domain, the finiteness of everything, the Earth's evolution and human limits.

The element carbon, finally, plays a central role as a basic component in life and the photosynthesis and a driver of global warming due to the huge anthropogenic emission of GHGs among which CO_2.

Its origin remained a mystery until the mid 20th century, when it was unraveled by the astronomer Fred Hoyle — after decades of research — who reached the conclusion that *carbon* originated during the explosion of giant stars in an astonishing sequence of coincidences.

The Big Bang

The wording "Big Bang" was coined by Fred Hoyle during a radio broadcast[12]. According to the inflation theory, a rapid exponential expansion of the initial "singularity" led to the present observable universe. If we imagine going back in time several billion years, we find a singularity, i.e. an infinite density point, with an immense temperature and pressure at a finite time in the past.

The process, Figure 2.1, is like viewing a movie backwards, from the end to the beginning. About 13.7 billion years ago, when the Universe is thought to have been concentrated in a singularity, a sudden expansion took place and continued for billions of years. The associated progressive cooling made the aggregation of matter possible, leading to the formation of big cosmic objects named galaxies, which in time split into stars and other bodies.

---

12 Fred Hoyle rejected the Big Bang theory (known as the standard cosmological model) that the Universe originated from a singularity and proposed instead a stationary model. The steady state model, proposed in 1948 by Hoyle, Gold and Bondi, holds that new matter is continuously created while the Universe expands.

The birth of the Universe, its expansion and associated phenomena are still under study, thus their description is limited to a few relevant aspects: expansion, the cooling of energy and matter, and the decrease in the internal energy of particles which allowed the formation of chemical elements[13] as we know them today.

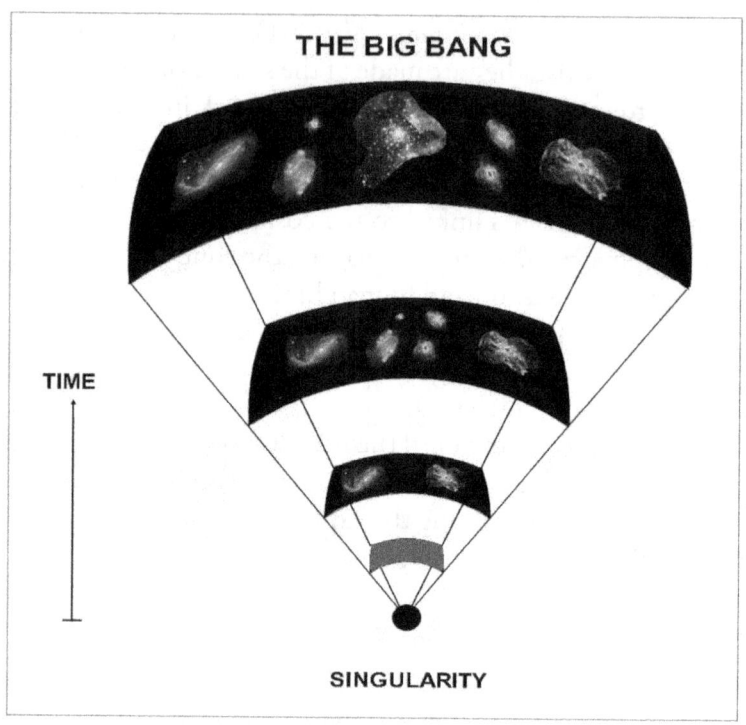

Figure 2.1 – The Big Bang and the cosmic expansion. Temperatures decrease with time and the Universe becomes larger and cooler.

When the Big Bang took place, the singularity underwent cosmic expansion, the four forces of nature — gravity, electromagnetism, strong and weak nuclear interactions — being one single force. The separation of gravity took place at 10^{-47} seconds after the Big Bang; from 10^{-36} to 10^{-32} sec, quarks, anti-quarks and electrons formed. Between 10^{-32} and 10^{-5} sec, the weak and strong forces and the electromagnetic force separated, followed by the formation of the first atomic nuclei, neutrons and protons.

---

13   The chemical elements were formed in two different phases named nucleosynthesis: the first resulted in the formation of Hydrogen, Helium and small quantities of Lithium and Beryllium, and the second, inside stars, generated the other elements represented in the Mendeleev Table. Source: abyss.uoregon.edu/

One second after the Big Bang temperature decreased by a few billion degrees Kelvin[14] (K) and energy and matter separated. When the Universe was about 200 seconds old, nuclei of Hydrogen and Helium began to appear, this most significant time being known as the first nucleosynthesis.

After about 15 minutes, stable atoms started to form, the Universe became transparent to light and radiation moved freely throughout the cosmos, making it visible. Around 200 million years after the Big Bang, proto-galaxies and quasars appeared, followed by the first galaxies between 500 million and one billion years ago. After two billion years, groups of galaxies emerged, and inside them stars formed between two and five billion years after the Big Bang. At that time, the second nucleosynthesis stage inside big and giant stars took place due to nuclear reactions. When the most massive stars of the first generation exploded, part of the matter cooked inside those stars was ejected into the surrounding space and then a second generation of stars formed, eventually giving rise to solar-like systems.

The Belgian mathematician George Lemaitre formulated the idea that the Universe could derive from a state of high density, using the concept of the "primitive atom" to describe it. The expansion from a high-density source model was confirmed (i) by the discovery of the recession of the galaxies (Hubble), (ii) by the recognition of cosmic microwave background radiation (Penzias and Wilson) and, (iii) by the finding that the proportion of Hydrogen and Helium has the same average value in all directions of space in which it is measured.

The recession was first discovered by Hubble who, after studying the growing distance between galaxies, proposed a constant value for the speed at which galaxies should have receded from the high-density singularity. Hubble's vision led to the conclusion that the universe is some 13.7 billion years old. The formation of the solar system is thought to have taken place 4.5 billion years ago when the universe was 9.2 billion years old, that is much smaller than today.

The second discovery is that of cosmic background microwave radiation. In 1965, Penzias and Wilson, following a series of measurements, confirmed that cosmic space is filled with electromagnetic radiation, the temperature of which is 2.7 degrees Kelvin (above absolute zero). They identified this cosmic radiation (or noise) as a residue from the time when the temperature of the universe decreased and stable atoms formed, being no longer destroyed by collisions with photons.

---

14   Zero kelvins (K) is the absolute zero, that is the temperature at which molecular motion stops. In terms of Celsius or Centigrade units 0 °C = 273.15 K, which is the melting point of water.

Cooling during about 13.7 billion years of expansion brought the photon radiation to its current temperature, as measured by Penzias and Wilson in 1965. The third finding in favour of Big Bang is that the measured abundance of elements is in agreement with the theoretical abundance: 75% Hydrogen, 24% Helium and about 1% of other elements.

During the 1960s, it was confirmed that the quantity of Helium of the galaxies in the Universe is equivalent to 24% of its total mass. This constant distribution of Helium, which is a light element, and its abundance second only to Hydrogen, brought to the conclusion that it originated together with Hydrogen within minutes of the Big Bang.

All other chemical elements of the universe — about 1 percent — were generated much later within the stars. The Big Bang, contrary to popular belief, was not an explosion, but rather an extremely rapid expansion of space combined with a simultaneous decrease in temperatures, due to which galaxies, stars and then solar-like systems formed.

Cosmological Constants and the Anthropic Principle

Nature's constants confer to the Universe a distinct character in terms of its structure and the events that unfolded. On the one hand, minimal variations in the value of the constants would have changed cosmic history, modifying the evolutionary path of the universe. Based on the values of these constants, a sequence of events from the Big Bang until today was formulated, providing a comprehensible, although not conclusive, picture.

The four laws of nature (gravity, electromagnetism, strong and weak interactions) are essential in understanding the Universe, cosmological "constants" determining its uniformity and symmetry. The initial conditions of the Universe, the differentiation within the proto-galactic cloud, the formation of galaxies and the evolution towards diversity and complexity fall into the field of thermodynamics, which controls the becoming of the cosmos throughout time.

The delicate balance of the constants that guided the development of the Universe and allowed events to unfold as they have done and the connections between the various branches of physics are surprising. The values of some constants are such that, had they been anywhere above or below these figures, the evolution of the Universe would have been very different and we would not have had a chance to be here.

The concept of the strong anthropic principle, which states that the Universe must be such to admit the presence of observers at a certain stage of its development, derives from Brandon Carter's[15] observations on universal constants.

This concept implies the existence of one single possible universe. Other scientists believe that multiple universes (multiverse) with other values of constants may exist. The Universe, cosmic objects and their evolution have been understood in the light of the invariant relationships among the cosmological constants.

The methodology of the anthropic principle, based on the values of the constants, was used by Fred Hoyle[16] to explain the formation of carbon within the stars, through the interaction of Beryllium. He also posed the problem to what degree the values of these constants were indispensable for the existence of us as observers.

Had values relative to nuclear and electromagnetic forces been slightly different from current values, the atoms of carbon could not have formed and life would not have originated as we observe it today.

The Formation of Chemical Elements

Mendeleev table, origin of chemical elements and the evolution of stars

Even before the discovery of sub-atomic particles and the formulation of a coherent theory on the atomic structure, some characteristics concerning chemical elements had already been observed. Among these was the frequency of their properties varying as a function of their mass.

Born in Siberia (1834-1907), Dmitri Mendeleev began arranging in the 1860s the 63 then known elements on the basis of atomic weights into groups sharing similar properties. In 1869, he published the Periodic Table, a scientific event comparable to Darwin's Theory of Evolution.

He organised the 63 elements according to their atomic weight, grouping those with similar properties in the same column and leaving some blanks for elements that were still unknown. During his lifetime *gallium, germanium* and *scandium* were discovered.

---

15 Brandon Carter (1942) is an Australian theoretical physicist, well known for his studies on black holes and the Anthropic Principle.
16 Fred Hoyle (1915-2001), English astronomer and mathematician.

His intuition, with the limited understanding of the atom at that time, was impressive as the elements discovered later did in fact have the properties he had imagined.

Figure 2.2 illustrates today's vision of the table, enriched by a series of discoveries during the latter part of the 19th and throughout the 20th century. The discoveries made between 1930 and the 1960s enabled scientists to unravel the complex story of the origin of elements.

The search for the basic constituents of matter was already advanced at the beginning of the last century, thus the existence of particles smaller than elements had been conceived, but the basic components of matter were still unknown.

Periodic Table of the Elements

Group	1	2	3	4	5	6	7	8	9	10	11	12	13	14	15	16	17	18
Period																		
1	1 H																	2 He
2	3 Li	4 Be											5 B	6 C	7 N	8 O	9 F	10 Ne
3	11 Na	12 Mg											13 Al	14 Si	15 P	16 S	17 Cl	18 Ar
4	19 K	20 Ca	21 Sc	22 Ti	23 V	24 Cr	25 Mn	26 Fe	27 Co	28 Ni	29 Cu	30 Zn	31 Ga	32 Ge	33 As	34 Se	35 Br	36 Kr
5	37 Rb	38 Sr	39 Y	40 Zr	41 Nb	42 Mo	43 Tc	44 Ru	45 Rh	46 Pd	47 Ag	48 Cd	49 In	50 Sn	51 Sb	52 Te	53 I	54 Xe
6	55 Cs	56 Ba	*	72 Hf	73 Ta	74 W	75 Re	76 Os	77 Ir	78 Pt	79 Au	80 Hg	81 Tl	82 Pb	83 Bi	84 Po	85 At	86 Rn
7	87 Fr	88 Ra	**	104 Rf	105 Db	106 Sg	107 Bh	108 Hs	109 Mt	110 Uun	111 Uuu	112 Uub	113	114 Uuq	115	116 Uuh	117	118

*Lanthanides	57 La	58 Ce	59 Pr	60 Nd	61 Pm	62 Sm	63 Eu	64 Gd	65 Tb	66 Dy	67 Ho	68 Er	69 Tm	70 Yb	71 Lu
**Actinides	89 Ac	90 Th	91 Pa	92 U	93 Np	94 Pu	95 Am	96 Cm	97 Bk	98 Cf	99 Es	100 Fm	101 Md	102 No	103 Lr

Figure 2.2 – Periodic table of chemical elements. The classification is based on the sequence of atomic numbers (the number of protons in the atomic nucleus). Rows are organized into columns (groups or families) whose elements have similar properties. Source: www.mendeleevtable.com.

By the 1930s it was clear that elementary particles such as quarks and leptons existed, that some atoms spontaneously underwent radioactive decay and that the abundance of elements in the Universe was more or less similar in all stars and similar to elements identified in the Solar System.

The peculiar position of Hydrogen, the first and the most abundant element in the Universe, was considered as an indication that all elements could derive from Hydrogen, a light element characterised by a nucleus made up of one proton and one electron.

Figure 2.3 shows the abundance of elements and star life cycles.

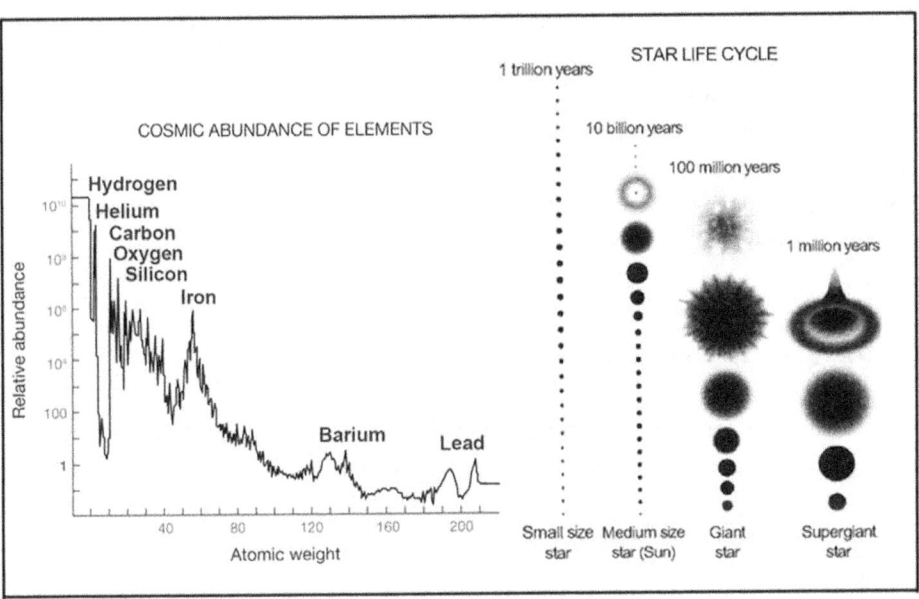

Figure 2.3 – (left) Abundance of chemical elements in the Solar System, (right) types of stars and their life span. Heavy elements form only inside giant and supergiant short-lived stars. Adapted form Martin Rees, Our "Cosmic Habitat", 2001.

As the most abundant in nature, Hydrogen accounts for 75% of the matter in the universe, followed by Helium, which accounts for 24%. Both were identified as components of stars.

The remaining 1% included all the other elements of the periodic system. By the late 1930s, Hans Bethe[17] and other scientists had identified components and phenomena inside stars: mass (mainly Hydrogen), temperature, pressure and energy flow.

By knowing these properties and the related nuclear reactions it was then possible to understand phenomena and products related to the life cycle of stars. Based

---

17 **Hans Bethe (1906–2005), German-American physicist best known for studies on the theory of stellar nucleosynthesis**

on this advanced knowledge, stars were broadly classified as shown in Figure 2.3 (right), which illustrates the evolution of different types of stars and helps to understand their life cycles based on their size. Those with a small mass burn like candles for an extremely long time (1 trillion years) and do not explode. Stars of medium size like the Sun have a life cycle of about 10 billion years, while giant and supergiant stars have much shorter life spans, in the range of 100 to 1 million years, ending with an explosion (observed as a supernova).

Major steps between the Big Bang and the formation of the Earth/Biosphere system are: (i) the origin of energy, matter, space and time, (ii) the formation of galaxies and stars, (iii) the formation of heavy elements in the core of giant and supergiant stars and the dispersion of dust clouds during their violent explosion, (iv) the formation of solar-like planetary systems by the compression of pre-existing dust clouds.

While this picture was generally accepted in the 1930s, the formation of heavy elements, among which carbon, was still not sufficiently clear.

The origin of carbon

Based on the finding that stars began forming about 1–2 billion years after the Big Bang, in 1946 Fred Hoyle formulated the hypothesis that heavy elements originated inside stellar kilns during the nuclear fusion of Hydrogen, and that Carbon resulted from the combination of three Helium nuclei. Thus two phases of nucleosynthesis were hypothesized:

- the first, 200 seconds after the Big Bang, during which nuclei of Hydrogen, Deuterium (Hydrogen isotope) and Helium formed from elementary particles.

- the second, tied to the life cycle of giant stars, after the Universe had expanded for one–two billion years, when galaxies and stars within them began to form. The combustion of Hydrogen and Helium within "furnaces" of giant and supergiant stars was thought to have given birth to heavy, second-generation elements, ranging from Beryllium to Uranium but the sequence of reactions was not clear yet.

The process of carbon formation was thought to happen in two steps: the collision of two Helium nuclei forming one Beryllium nucleus, which then would collide with another Helium nucleus to generate Carbon. However, because Beryllium nuclei are highly unstable and decay instantaneously, it was highly improbable that Beryllium could capture a nucleus of Helium and produce Carbon in the quantity find in nature.

In 1957 Hoyle, Fowler, and Geoffrey and Margaret Burbidge[18] argued that an encounter between Beryllium and Helium could only take place under different conditions: when the inside of a star runs out of for Hydrogen to fuse, its core starts collapsing until the temperature rises to 100 million Kelvin.

At that temperature and pressure, Helium nuclei can combine at a sufficiently fast rate to rival the speed of Beryllium decay back to Helium, so that enough Beryllium–Helium collisions can take place generating Carbon, which is then a very stable element. The high probability that a collision between Beryllium and Helium will produce Carbon lies in the similar energies of certain resonances (excited states) of the Helium, Beryllium and Carbon atoms.

The formation of heavy elements was therefore definitely tied to the final explosion of stars of great dimensions. It was discovered in sum that Carbon and other heavy elements originated only at the extremely high temperatures available in the final stage of giant and supergiant stars.

According to Martin Rees, the complex origin of Carbon indicates that our universe is based on laws surprisingly favourable to life, which is made out of C, H, O, N, P and S, considered the "building bricks" of the biological world.

---

18   Hoyle, Fowler, Geoffrey and Margaret Burbridge were the co-author scientists of the paper B2FH (initials of their surnames) on stellar nucleosynthesis.

CHAPTER 3

The Earth and The Biosphere

Solar System, Planets, Asteroids and Comets

The present knowledge of the Solar System, shown in Figure 3.1, is largely based on observations made through telescopes, satellites and space flights during the past century.

According to the nebular hypothesis, the formation of the Solar System 4.5 billion years ago was most probably triggered by the explosion of a supernova[19] and the associated compression of a pre-existing *gaseous nebula*, which underwent a gravitational collapse. Pressure waves caused the contraction of the nebula which flattened into a rotating planetary dust disc, with a stellar mass at its centre.

---

19   A supernova is a type of star that explodes, becoming extremely luminous and causing a burst of radiation because of nuclear fusion. The compression of a pre-existing gas and dust cloud can start the formation of a planetary system. The first explosion of a supernova was observed from China in 1054 A.D.

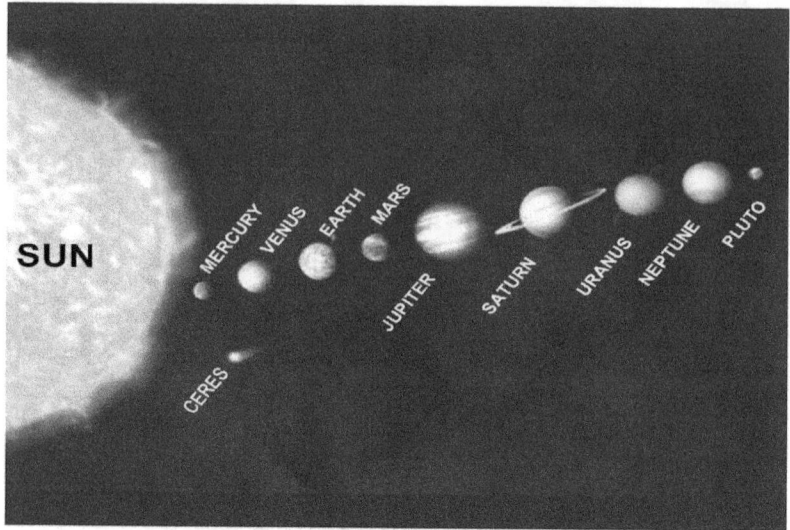

Figure 3.1 – The Solar System. From left: the Sun, Mercury, Venus, Earth, Mars, Jupiter, Saturn, Uranus, Neptune, Pluto (distances not to size).

The Sun is the parent star of the system, its mass and density being sufficiently high, enough to support nuclear fusion. The huge quantity of energy radiated into space consists of electromagnetic waves (including visible light) responsible for the warming of planets. The special condition of our planet is that the distance from the Sun is just enough to support the presence of a biosphere on Earth, which is thus located in the life habitable zone of the system. The planets as we know them today originated from the dust and gaseous substances in the disc, at the time shaped somewhat like Saturn's rings. Their formation was due to the accretion of dust particles initially into small lumps, which later attracted to one another, creating bodies of various dimensions, probably up to several kilometres in diameter. Their collisions led to the formation of bigger masses named *planetesimals* — which became progressively larger due to the gravitational attraction of smaller bodies — finally reaching the size of planets. Those closer to the Sun — made of metals and rocky silicates with high melting points[20] — transformed into the terrestrial planets Mercury, Venus, Earth and Mars. The more volatile gaseous substances originated the outer gas giant planets Jupiter, Saturn, Uranus and Neptune.

The selection process, that led to the present distribution of planets into an internal and an external zone, was driven by the Sun's force of gravity. The temperature distribution in the dust disc (higher in the internal zone and progressively lower

---

20 Metals and silicates in a solid or liquid state can stand a higher temperature than gaseous planets (such as Jupiter and Saturn).

in the outermost area), and the innumerable collisions, that caused the clustering of masses in a selective order of dimensions, gave birth to the planets as we know them today. During the accretion process it is believed that Jupiter (the Solar System's largest planet) disturbed the clustering of the small masses of different dimensions between its orbit and that of Mars, thus originating the asteroid belt. When the Hydrogen fusion process inside the Sun started, a solar wind, at a velocity of 2 million kilometres per hour, began expelling the smallest dust particles which were still present, stabilizing the growth of the inner rocky planets. The Sun — with a life cycle of 10 billion years half of which have already passed — is expected to end its journey as a giant red star that will merge with surrounding planets. About 300 million years after the formation of the Solar System, the asteroid bombardment ended and the system acquired its present shape with planets spaced as they are now. Only a very small portion of the Sun's energy is absorbed by the Earth's biosphere and a much smaller amount is used in photosynthesis.

Mercury, the closest to the Sun but smaller than the Earth, has a very thin atmosphere, a surface marked by impact craters and temperatures reaching 400°C.

Venus is believed to have had great oceans in the past, which evaporated forming a layer of clouds that turned into a very dense atmosphere. The planet lacks liquid water and its surface temperatures reach 457°C, due to a giant greenhouse effect. Strong winds move its clouds at a much higher speed than that of the rotation of the planet. The terrestrial atmosphere, by comparison, moves at the same speed of the planet and this makes the Earth hospitable to life. Venus has no magnetic field, but an enormous iron nucleus and surface forms typical of volcanic activity, lava flows covering 85% of its surface.

Earth shows a clear evidence of significant geological activity which has persisted since the formation of the planet, heavily affecting the evolution of the biosphere. Earth holds today a limited number of impact craters and is characterized by intense surface erosion, volcanism and tectonics which have wiped out most traces of impacts. The 4 billion years old biosphere has been protected from the effects of solar wind by the Earth's magnetic field. Solar radiation, atmosphere, hydrosphere, lithosphere and life interact within the biosphere, making photosynthesis possible, a process through which solar energy is transformed into chemical energy, which is in turn the basis of living organisms. The uniqueness of the Earth and biosphere, compared to other planets, is due to the distance from the Sun, orbital properties, diameter, mass and density, chemical composition and the distribution of different materials (core and mantle and the energy they contain).

Mars, known as the red planet, is the most similar to Earth, has an iron nucleus and a crust of silicates, it is internally active and has great volcanic craters and polar ice caps. Its scarce amounts of Oxygen, an extremely rarefied atmosphere of CO_2 and mainly the distance from the Sun are the reasons for the very frigid surface temperatures. Mars's surface is frozen, impact craters are virtually absent and its red colour is due to sheets of sand. Information from the Mars Exploration Rover showed erosion morphologies shaped by running water.

The *asteroid belt*, located between Mars and Jupiter, is composed of rocky bodies with diameters from a few metres to 1,000 kilometres. The closer the asteroids are to Mars, the greater the possibility that their orbits will be disturbed and deviated towards the inner rocky planets. The probability that an asteroid with a diameter of 1 km collides with the Earth is one in a million years. Asteroids of around 50 m have higher probabilities of once in a thousand years, and 10 to 5 m objects disintegrate in the atmosphere. The Permian and the Cretaceous mass extinctions, respectively occurred 240 and 65 million years ago, are attributed to asteroid impacts.

The external planets *Jupiter and Saturn* are vast gaseous bodies made of Hydrogen and Helium, with masses respectively 318 and 95 times that of the Earth. If Jupiter's mass had been 10 times larger, the related high pressure and temperature would have triggered nuclear reactions leading to the formation of a small star greatly influencing the future assemblage of the Solar System. Jupiter's atmosphere is made up of 75% Hydrogen and 24% Helium and 1% other elements.

Uranus and Neptune are respectively 14 and 17 times larger than the Earth. They have smaller amounts of hydrogen and helium in their atmospheres compared to the giant planets and contain great quantities of water, ammonia and methane in the form of ice.

Comets are frozen dirty snowballs that orbit around the Sun, their pathways being easily affected by celestial bodies passing by. Comets are made of rocky material, ice, dust, carbon dioxide, methane and ammonia. When passing near the Sun they can be vaporized and their tails fragmented. The comet Hale-Bopp was seen for several weeks in April 1997 at a distance of 198,000 km from Earth. The comet Shoemaker-Levy that orbited around Jupiter collided with this giant planet in July 1997, images being sent in real time by the NASA spacecraft Galileo.

The *Oort Cloud* (probably expelled from the Solar System upon its formation) has a spherical shape and consists of trillions of frozen bodies (comets) surrounding the Solar System at a distance of 50 to 100,000 AU[21]. The attraction from the

---

21   1 Astronomic Unit is equal to 149.58 million kilometres, the average distance between the Earth and the Sun

Sun is very weak, however the orbits of the Oort Cloud comets can be disturbed by other factors (eccentricity and combination of planets' orbits) that can divert some of these ice bodies into the Solar System.

The Earth's Geological and Biological Evolution

Between 4.5 and 4.2 billion years ago, asteroid impacts on the newly born Earth helped maintain both the planet's surface and its interior in a liquid state. The heavy metals, mainly iron and nickel, migrated towards the core of the planet and the light magmas were displaced outwards, forming the mantle and crust. The Earth's internal structure, the motion of crustal plates, earthquakes and volcanic activities and the associated emission of water vapour, greatly influenced the formation and the evolution of the biosphere. The heat flow from inside the Earth to the surface — which amounts today to 31.1 TW per year — is due to the thermal energy transfer by conduction and convection[22]. Convection concerns both isolated volcanic eruptions under the oceans and on land and the huge continuous sub-marine magmatic emissions along the oceanic fractures of the Mid-Atlantic and the East Pacific Ridges[23]. Earth receives energy from the Sun, its interior and more recently from human activities. The formation of the Earth's crust began 4.2 billion years ago, when meteorite impacts diminished in frequency. This deduction was derived through the study of zircons[24] 4 billion year old rocks in the Jack Hills (Western Australia). Figure 3.2 shows the Earth's internal structure and the last 225 million years migration of crustal plates.

The Earth has a solid inner core, containing mainly iron, while the outer liquid is the source of the magnetic field through the magmatic convection. Above the core is the mantle upon which the solid crust floats. Temperatures decrease from 6000°C in the solid core to 3700°C between the liquid core and the mantle, and keep decreasing upwards to the surface temperature of the crust.

---

22 Conduction is the transfer of thermal energy form a higher temperature region to a lower temperature zone in direct contact, without flow of material. Convection is the transfer of thermal energy related to sub-crustal or mantle movement of liquid and gaseous materials, as it occurs during volcanic eruptions on emerged land and the sea floor (mainly along the Mid-Atlantic Ridge and the East Pacific Rise).

23 Mid Atlantic and East Pacific fractures are due to mid-oceanic divergent tectonic plate boundaries (see Figure 3.3).

24 Zircons are the oldest and almost indestructible crystals on Earth.

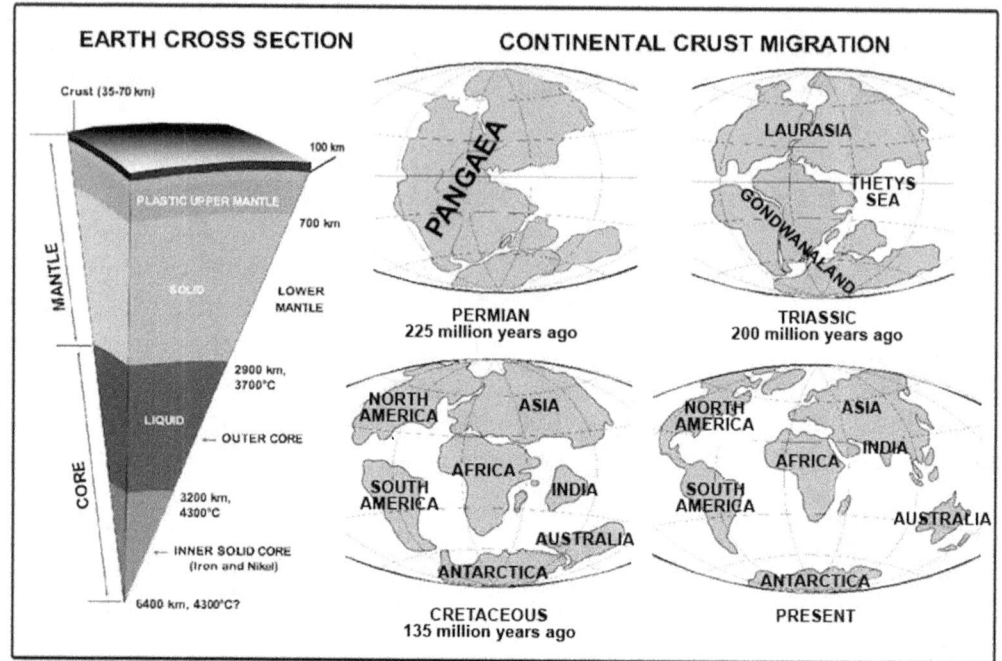

Figure 3.2. (left) Cross section of the Earth. (right) The migration (or drift) of plates from the super-continent Pangaea. The Thetis Sea formed in the Triassic and from the Cretaceous to present plates separated further. A detailed set of "Continental Drift Maps" is available in: www.adonline.id.au

The density of the Earth's materials (estimated through the velocity of seismic waves) decreases in an outward direction, ranging from 12 to 8 T/m^3 in the core and from 7 to about 4 T/m^3 in the mantle. The solid crust on which we live (35 to 70 km thick) has an average density of 2.7 T/m^3 and floats on the plastic magma of the upper mantle. Convective movements of magma drive the migration and the break-up of plates[25] forming the crust, the subduction of the oceanic floor, the formation of volcanic arcs and mountain chains, the emission of lava along the Pacific Ring of Fire, the East Pacific and Mid Atlantic fractures, as shown in Figure 3.3.
The bottom graph shows the cross section of the Nazca plate, the formation of the ocean floor on both sides of East Pacific fracture, due to the continuous emission of magma, and the subduction below the South American continent. well-known case of plate tectonics is the collision of India with the Euro-Asian continent (a 7,300 km northward movement that lasted 90 million years),

---

25 The lithosphere is broken into several plates, the long-term migration of which is shown in Figure 3.2

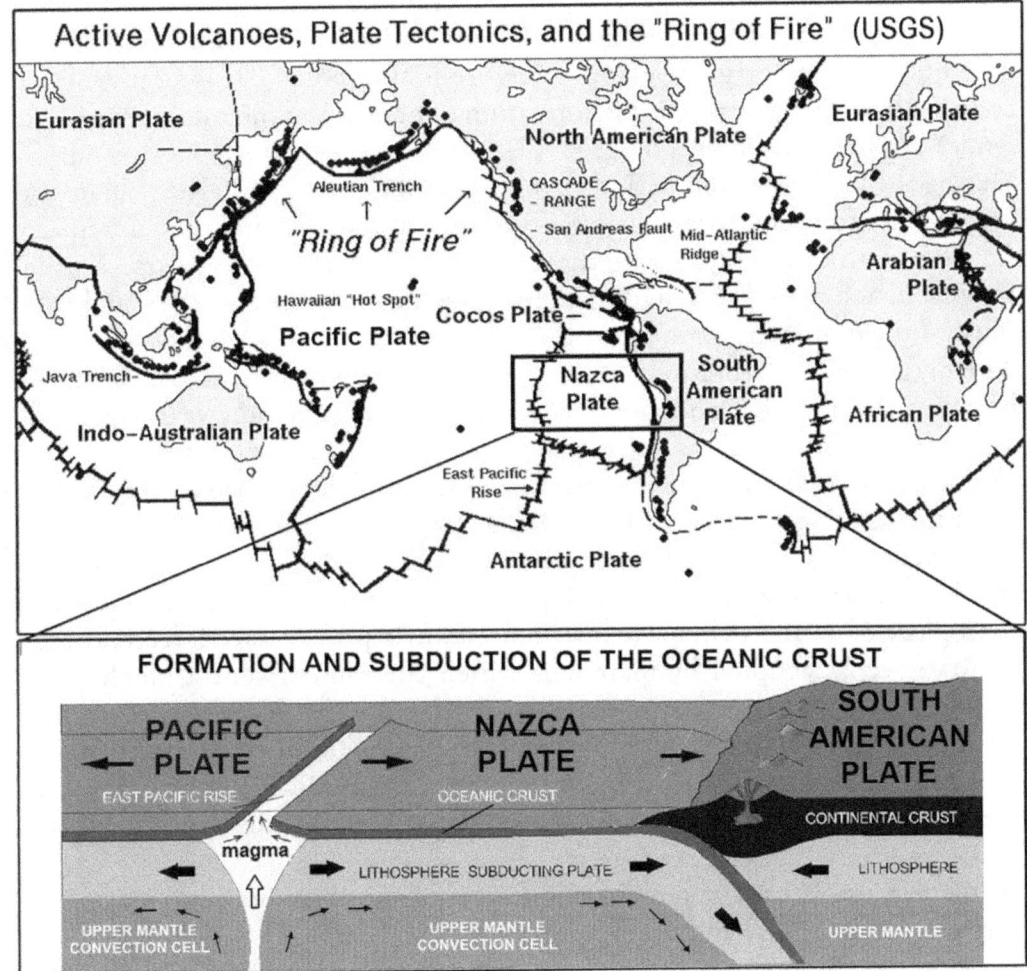

Figure 3.3 – (top) Map of Tectonic Plates, the Pacific Ring of Fire and the oceanic fractures East Pacific Rise and the Mid-Atlantic Ridge. (bottom) Cross section of the Pacific Ocean crust and the subduction of the Nazca Plate below the South American plate. Sources: (top) USGS Topinka 1997, modified after Hamilton 1976, Tilling, Heliker and Wright 1987, oceanexplorer.noaa.gov, (bottom) oceanic crust formation and subduction, adapted from pubs.usgs.gov

leading 55 to 49 million years ago to the formation in the Early Cenozoic (see Chapter 5) of the Himalayan Range and the Tibetan Plateau. The superficial portion of the crust on which we live is made up of a collection of fragments resulting from the interaction of past geologic events (plate tectonics, surface erosion, volcanism, earthquakes). When continents separated, life underwent a diversified evolution: organisms once living in South America and Africa (formerly part of Pangaea) evolved differently, as documented by their fossils. The present assemblage of plates therefore is the result of a long journey of continents which gave rise to the formation of oceans and mountain ranges like Andes and the Himalayas.

During the Carboniferous period (340–290 million years ago), the climate was very hot and humid, forests covered the planet and the largest coal deposits formed in this period. When the Carboniferous ended, the first reptiles populated and then dominated the Earth together with other species, and from then on the biosphere swarmed with terrestrial life.

The biggest mass extinction, which wiped out 96% of marine life, occurred between the Permian and the Triassic periods 251 million years ago. The dimension of the crisis was due to a meteorite impact (the crater was identified in Antarctica) which caused earthquakes, huge volcanism, release of CH_4 from the sea floor and climate change. At the end of the Permian period — coincidentally close to the biggest mass extinction — big regional eruptions of basalts[26] occurred worldwide. During the most recent mass extinction, which took place 65 million years ago, the majority of reptiles disappeared and their habitats were occupied by mammals which then underwent a much faster evolution.

Pre-hominids appeared 5 million years ago. Figure 3.4 shows, from top, Eons (the great divisions of the geological time scale), the Phanerozoic[27] Eon, the Cenozoic Era and the evolution of biodiversity during the last 542 Million years.

In the top graph the oxygen enrichment sequence of events (essential to the formation of biosphere and the evolution of life) is shown:

- from about 4.5 to 2.45 billion years ago the atmosphere did not contain oxygen or in very small quantities

---

26 Huge eruptions of basalts occurred at the end of the Permian in Siberia, Deccan (India), Karroo (South Africa), Ethiopia, Rio Parana region in South America, and along the eastern coast of the United States.

27 Phanerozoic: the term is derived from Greek and identifies the time in which evidence of life became visually clear through the presence of fossils in rocks. The geologic history of the Earth (Figure 3.4) is divided in 4 eons (grand divisions of geologic time): Archean, Archeozoic, Proterozoic and Phanerozoic.

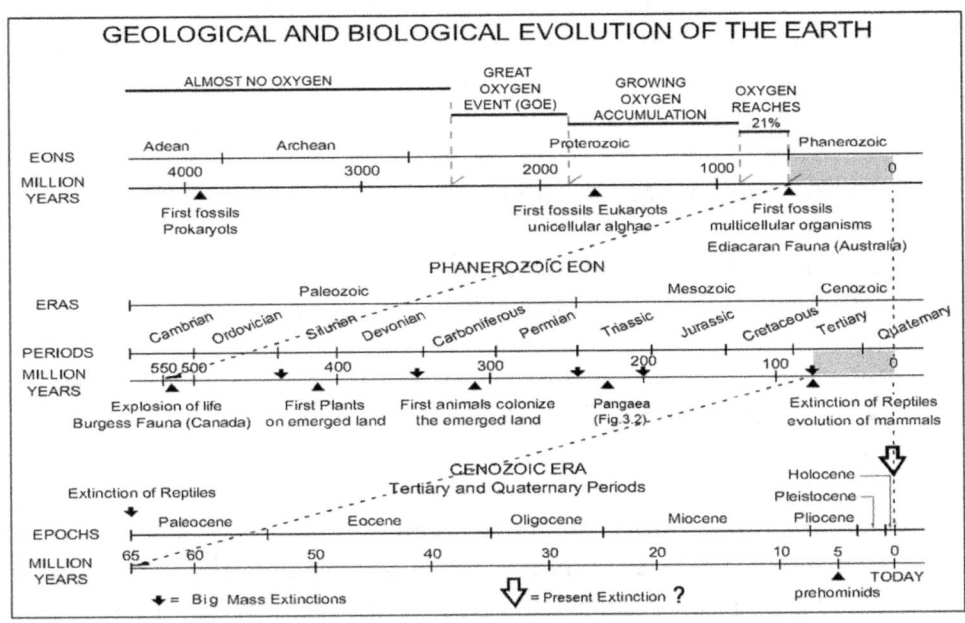
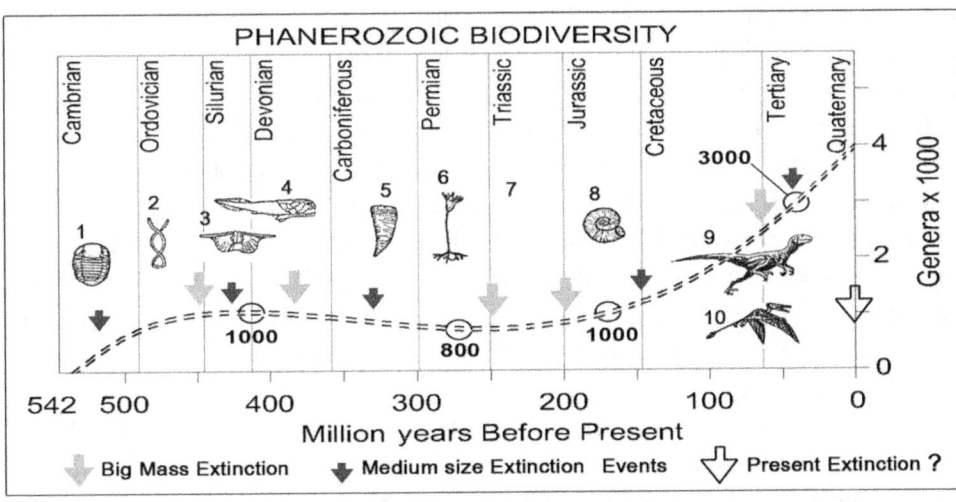

Figure 3.4 – (top) The oxygen enrichment of the atmosphere, the geological and biological history of the Earth and major events. The figure is adapted from E.O. Wilson, "The Diversity of Life", 1992, Cambridge Mass: The Belknap Press of Harvard University Press. (bottom) Phanerozoic diversity trend of life and mass extinctions. (Fossils: 1 trilobite, 2 graptolite, 3 brachiopod, 4 fish, 5 belemnite, 6 crinoids, 7 concerns the biggest (Permo-Trias) mass extinction (which included marine life 97%, terrestrial vertebrates 70% and insects 87%), 8 ammonite, 9 and 10 reptiles (dinosaur and pterosaur). Data on biodiversity trend are from Sepkoski J.J., Miller A.I., in Phanerozoic Diversity Pattern *(Profiles in Macroevolution).*

- from about 2.45 to 1.85 billion years ago a catastrophic enrichment occurred (the Great Oxygen Event, GOE): the oxygen released in great quantities by microorganisms was mainly absorbed by ocean water and oceanic floor sediments (from which the Banded Iron Formations originated); then in the absence of further amounts of oxidizable minerals oxygen started accumulating and enriching the atmosphere

- from 1.85 to 0.85 billion years ago, oxygen was growingly emitted by the oceans and absorbed by emerged land and the ozone layer formed

- from 850 to 700 million years ago, oxygen rapidly enriched the atmosphere which reached its actual composition in volume with 21% oxygen, 78% nitrogen and 1% carbon dioxide and argon.

The Ediacaran fauna in Australia — dated late pre-Cambrian, 580 million years ago — includes the first stationary multi-cellular, soft-bodied and frond-shaped organisms. The Cambrian explosion of marine life occurred around 542 million years ago is documented by the Burger Shales in Canada which hold an abundant fossil fauna. The formation of the ozone layer (essential in protecting life from ultraviolet radiation) was completed around 400 million years ago, the time at which the land began to be colonized by marine vegetation, followed 310 million years ago by the tetrapods[28]. About 240 million years ago, when the proportion of O_2 in the atmosphere reached the level of 21% in volume, the formation of the ozone layer was completed and animal life populated the continents. The trend of biodiversity (Figure 3.4, bottom), expressed in thousands of genera[29], shows an average value of 1,000 genera from the Silurian to the end of Jura, then followed from the beginning of the Cretaceous (145 Million Years ago) by a nearly constant increase to the current value of 4,000 genera.

A sudden extinction occurred 65 million years ago (K-T boundary) due to a meteorite impact, after which life recovered and the increase of biodiversity matched with a growing cooler climate (see Figure 5.3).

---

28 Tetrapods are vertebrate animals with four limbs (feet, hands, wings). Amphibians, reptiles (limbless tetrapods by descent), dinosaurs, birds and mammals are all tetrapods.

29 Biological classification is hierarchical, from the three great Domains (Archaea, Bacteria and Eukarya), through Kingdoms, Phyla, Classes, Orders, Families, to Genera, which are groups of species with common characteristics, including tens of millions of species (among which is *Homo sapiens*).

The Earth's Magnetic Field

The Earth's magnetic field, Figure 3.5, is the shield against solar wind and thus the special coincidence which made our planet suitable for life. Mars, which is farther away from the Sun compared to the Earth, probably lost its atmosphere due to the absence of a magnetic field and weaker gravity. The magnetic field is generated in the liquid outer core (see Figure 3.2) by a self-exciting dynamo process. Convective motions inside the electrically conducting fluid of the outer core, interact with Coriolis forces due to the Earth's rotation and produce a dynamo-like magnetic field in the absence of which atmosphere would have been dispersed into space long ago.

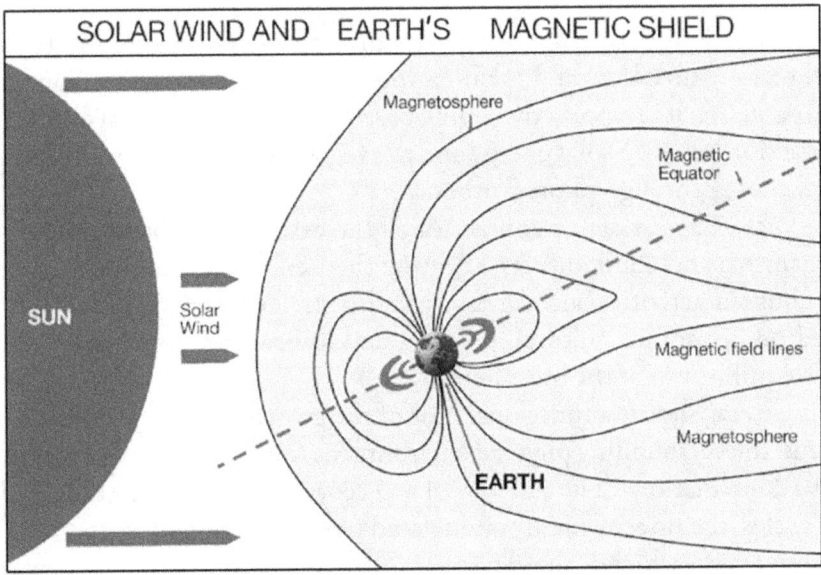

Figure 3.5 – Solar wind and the Earth's magnetic field.

The Earth's magnetic field began developing about 100 million years after the formation of the Sun which started acting as a shield deviating solar wind, blowing at a speed of more than 2 million kilometres per hour, into space.

Important terrestrial phenomena are the variation of the magnetic field, the migration of poles and the polarity inversion[30] which occurs on a time scale of

---

30 The polarity inversions (the most recent one occurred some 800,000 years ago) have been identified thanks to the variable orientation of magnetite crystals that formed during the cooling of the magma ejected along the Atlantic and the Pacific ocean floor fractures.

tens of thousands years most probably due to heat fluctuations in the liquid outer core.

The Earth's magnetic field has been essential to navigation, to the orientation of people on land and to the migration of animals on land, sea and in the air.

The Biosphere

Size, composition, environment, ecosystem and living organisms

The Biosphere is the spherical vault extending from approximately 10 km above sea level to 6–7 km below and includes four major interacting components: land, water, air and life. Its complex dynamics, based on bio-geochemical cycles, is due to the interaction between solar energy and the vegetation which transforms light into chemical energy through photosynthesis.

The *natural environment* consists of all the external conditions that can influence organisms and communities, affecting their existence and development. Very recently human activities and the associated complexity of the *social environment* — based on economic, cultural, political and ethical components — altered the dynamic equilibrium of the biosphere.

A single *ecosystem* is a functional unit of the environment and consists of living organisms, the surrounding physical environment (air, water, terrain and rocks) and the mutual interactions. The stability of an ecosystem depends on the regularity of natural cycles, the flow of energy and related transformations.

Homeostasis[31], which is the capacity of life to regulate its internal environment in order to preserve a stable condition, has been perturbed innumerable times during the last 4 billion years. Documented big mass, and a number of middle-size and local life extinctions occurred during the last 542 million years.

A recent study on the sudden extinction of large mammals in North America[32] was carried out by Eric Scott, a palaeontologist at San Bernardino County Museum in Redlands California. The Earth and the Biosphere are so tightly integrated that

---

31 Homeostasis is a system's property that regulates its internal environment maintaining a stable constant condition.

32 The synthesis of findings was reported in the Scientific American, March 2011, by Rebecca Coffey under the title "*Bison versus Mammoths*".

they are often referred to as a single system. Without the planet's astronomical and physical characteristics, the constant flow of solar energy and the protection from the magnetic field the biosphere would not exist.

Due to this complexity, the Earth and Biosphere can be considered an extraordinary cosmic laboratory in which matter and energy, natural laws and life are astonishingly blended and integrated. Water, the only chemical compound present on the Earth in all three states (solid, liquid, gaseous), is essential to living organisms which contain a large quantity of it.

Oceans represent the first environment on the Earth that protected life from ultraviolet rays for more than 3 billion years, a period during which emerged land was unfit for organisms. The oxygen enrichment[33] of the atmosphere and the formation of the ozone layer are essential phenomena in the biosphere's evolution. Ozone molecules O_3 in particular filter the ultraviolet radiation preventing it from reaching the surface of the Earth and being harmful to life.

The tree of life

The Biosphere is a very unique cradle of life in which all living organisms interact with the hydrosphere, atmosphere and lithosphere. Life's origins, its essence, complexity and perseverance in surviving during dramatic geological, environmental and climate changes have always been at the centre of scientific and philosophical reflections.

The scientist Carl Woese[34] proposed in 1977 a phylogenetic tree of life based on RNA[35], identifying three life domains: Archaea, Eubacteria and Protists or Eukaryota[36].

Figure 3.6 represents the 3 domains of life, from the first cell which appeared around 4 billion years ago on the Earth, probably brought by meteorites from the outer space.

The first two branches, Bacteria and Archaea, respectively include very old and more advanced prokaryotic microbes, the cells of which neither have a nucleus nor a membrane.

---

33 Source: Great Oxygenation Event (Wikipedia)

34 Carl R. Woese, Professor of Microbiology at the University of Illinois at Urbana Champaign

35 RNA, ribonucleic acid is an important molecule, made up of a chain of nucleotides, like the double-chain DNA (deoxyribonucleic acid), but different in its single-chain structure.

36 Archaea and Eubacteria are the oldest unicellular prokaryotic organisms which neither possess a nucleus nor membranes. Eukaryots are unicellular and multicellular organisms (metazoan), with cells based on a complex structure inside a membrane.

The third branch, Eukaryota, includes all the other forms of life (fungi, plants and animals) based on unicellular and multi-cellular organisms, their cells having a well-defined nucleus and a membrane.

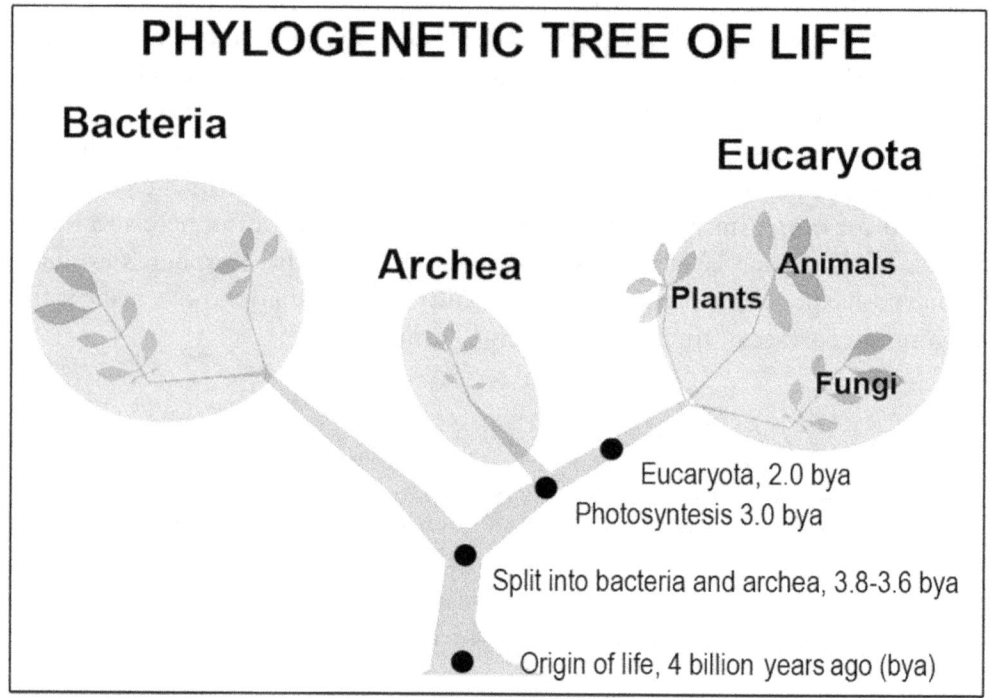

Figure 3.6 – Phylogenetic tree of life. The first cell appears 4.0 billion years ago (bya). Unicellular life splits into bacteria and archaea (Prokaryota) 3.8-3.6 bya. Photosynthesis begins 3.0 bya. Complex cells (Eucaryota) appear 2.0 bya, Multicellular life (metazoan) begins developing around 1.0 bya. The Ediacaran fauna (Adelaide, Australia) and the Burges Shales fossils (Canada) are respectively dated 0.580 and 0.520 bya.

According to Jay Gould[37], Burgess' fossilised remains consist of about 20 types (among which Chordates, the phylum[38] to which we belong to) different one from another and different from the life forms living today.

---

37 **S. Jay Gould (1941-2002) was the paleontologist who revised and reinterpreted the Burgess Shales fossils collected between 1909 and 1924 by C. D. Walcott.**
38 **Phylum is a major unit in the taxonomy of animals**

Requirements for the Formation of a Biosphere and the Rise of Intelligence

The search for life and intelligence outside the Solar System started about ten years ago with the Planetary Society's programme SETI[39] (Search for Extra-Terrestrial Intelligence) and is still ongoing. The search for Earth-like planets is based on the hypothesis that conditions similar to ours may exist in other solar-like systems.

Figure 3.7 shows the evolution of the universe from the Big Bang, represented through some major steps: galaxies, stars, the Solar System and the Earth which hosts life and more recently intelligence.

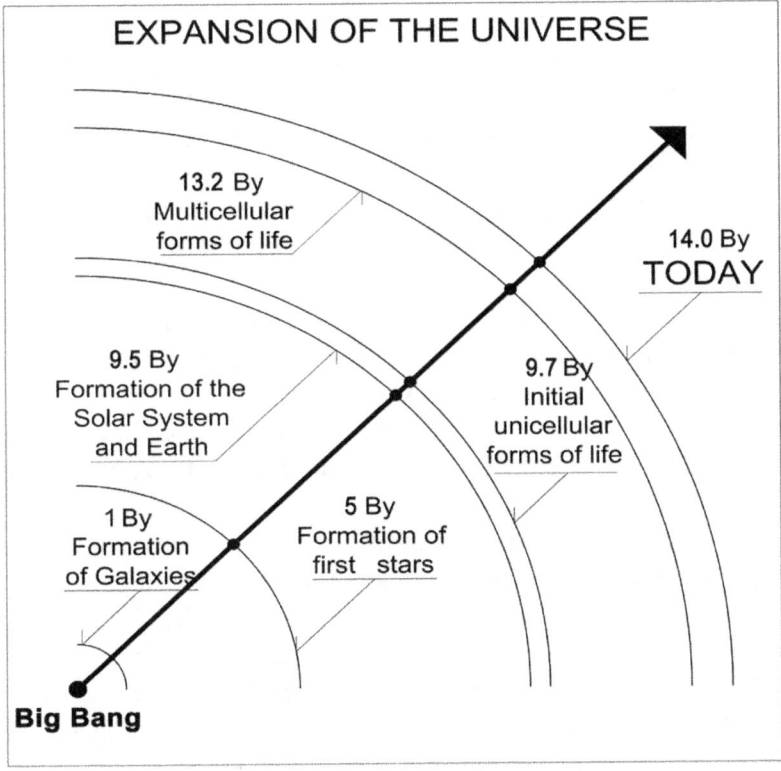

Figure 3.7 – An overview of the universe from the Big Bang (By: Billion years).

The thermodynamic death of the Sun will occur 5 billion years from now when the Universe will be around 19 billion years old.

---

39   Source: www.planetary.org

The formation of the terrestrial biosphere and the emergence of intelligent life needed specific cosmic, planetary, chemical, pre-biological and biological conditions. The attempt to imagine the requirements for the formation of a biosphere are summarised below, in line with the view by Joan Oró[40], the Spanish geochemist who significantly contributed to the understanding of the origin of life.

Her basic concept is that intelligent life would be virtually impossible elsewhere unless a sequence of phenomena, similar to the formation of our Solar System, the Earth and the Biosphere repeat in other planetary systems in a way that would allow intelligence to emerge.

The starting point is that the Universe is populated by 100 billion galaxies or so, each containing an order of magnitude of about 100 billion stars, and that the conditions allowing for the formation of life could be present in many other solar-like systems. A single, small to medium-sized star (not one belonging to a binary system) would be needed for the formation of sun-like planetary systems. This implies in turn a second or third generation star, deriving from a gas cloud expelled from a first generation big star during the final explosive stage.

The star's life cycle, similar to our sun, would need to be approximately 10 billion years, because the biological journey leading to the emergence of intelligence in our biosphere began around 4.2 billion years ago and intelligence developed during the last few million years only.

The greater the number of planets orbiting a sun-like star, the greater the probability that one of these planets may have an orbit neither too close to the sun like Mercury and Venus, nor too far like Mars. Important factors for a planet to be colonised by life are the size and internal makeup, the presence of a hydrosphere, an atmosphere, a lithosphere and elements including C, H, O, P, N and S (the bricks of life).

The geological activity of the earth-like planet is essential, as well as internal quality and distribution of matter that allow for the formation of a magnetic field protecting the newly born planet from the solar wind. The presence of water and organic compounds, as well as solar energy that reaches the planet at a temperature favouring the chemical reactions that take place on the Earth, are essential.

Also fundamental are the initial absence or shortage of oxygen and hydrogen as well as a neutral ph that favours the stability of chemical bonds.

---

40 J. Oró, (1999). Evolutionary requirements for the development of intelligent life from the book by Colombo, R. G. Giorello, E. Sindoni: *Origine della Vita Intelligente nell' Universo (Origin of Intelligent Life in the Universe)* International Conference, Villa Monastero (Lecco, Varenna, Italy). Edizioni Newpress, Como, Italy.

The following steps are identified in Orò's description:

- given the above limits, based on the assumption of 100 billion stars in the Milky Way, there could be some **2 million** planets similar to Earth.

- the chemical, biochemical and self-organizing stages of evolution need to lead to the formation of molecules and to self-replication. The formation of DNA and enzymes is essential in this process. Microfossils have been identified in rocks from about 4 billion years ago, and the presence of DNA and RNA suggest an already high level of evolution and complexity that can be explained by the evolved molecules probably deposited on Earth by comets from outer space. This would imply the origin of life elsewhere. However, the presence of too many comets would imply too many dangerous collisions, which could only be avoided by the presence of a big gaseous planet like Jupiter, (on which the comet Shoemaker-Levy impacted in July 1994). Based on this, planets with characteristics similar to the Earth with its biosphere narrows down to **1 million**.

- if we take into consideration the evolution from prokaryotes to eukaryotes, the switch to sexual reproduction and the evolution of the entire Biosphere, the estimation narrows down to **100 thousand planets** on which eukaryotes could live.

- since about 650 million years ago, a number of extinctions have occurred, the one across the Permian-Trias limit concluded with the extinction of 96% of marine species. The last extinction 65 million years ago wiped out the dinosaurs, which had dominated the Earth for 130 million years, allowing however mammals to emerge and occupy ecological niches from which they had been previously excluded. The estimated number of planets in the Milky Way on which intelligent organisms could be living then narrows to **10 thousand** because of extinctions.

- the rise of hominids 5 million years ago, the growing complexity of their nervous system, brain volume and capabilities, and the trend of the climate, languages, and very recently philosophy, arts and science, narrows the estimate to only **100 planets**.

The possibility of humans interacting with other intelligent organisms — no matter if there are 100 or 1,000 or 10,000 earth-like planets in the Milky Way — is unrealistic due to the distances. Moreover, a different evolutionary stage of intelligence on another planet could be an insuperable obstacle. This implies that we and future generations may never know anything about the existence of alien forms of life or intelligence due to distances between our and other twin solar systems.

The discovery of primitive and unicellular forms of life on Mars could represent a step forward, but not necessarily lead to a conclusion that life is abundant in the cosmos. The accidental and unpredictable sequence of events that occurred on Earth makes communication with other inhabited planets extremely improbable.

If the meteorite that hit the Earth 65 million years ago had been double or triple in size, it could have entirely eliminated the possibility of the rise of intelligence or might have postponed the present evolutionary stage, or finally made the presence of intelligent organisms impossible. After the Big Bang, matter and energy underwent a self-organizing process that evidently included the chance of intelligent life.

Today the possibility of the presence of the bricks of life (C, H, O, P, N, S) elsewhere in the cosmos is shared by the majority of scientists. The development of intelligence, on the other hand, as a consequence of unpredictable and innumerable coincidences during the evolution of life as depicted by Joan Oró, makes the formation of other biospheres, in a trend similar to the one on the Earth, an exceptional event.

NASA's Kepler spacecraft confirmed (from internet news, December 5, 2011) the discovery, outside solar system, of a habitable planet with a radius 2.4 times the size of the Earth's radius and an average surface temperature of 22°C.

CHAPTER 4

From Prehominids to Homo Sapiens

Prehominids and Hominids

Mammals, the class of vertebrates to which humanity belongs to, diverged from *sauropsids*[41] at the end of the Carboniferous period 300 million years ago. During the Permian, Jurassic and Cretaceous[42] periods (300 to 65 million years ago) mammals underwent slow changes, surviving to three extinction events (see Figure 3.5).

Emerged lands from 200 to 65 million years were mostly dominated by the explosion of the *sauropsids*, which branched into reptiles and birds. The impact of a meteorite 65 million years ago at the Cretaceous-Tertiary boundary, caused the last big mass extinction and the scene abruptly changed. Great reptiles (dinosaurs)

---

41 *Sauropsids* are a group of vertebrates including extinct species and reptiles present today.
42 See Figures 3.4.and 3.5.

which had dominated for about 130 million years were wiped out while moderately affected were the small mammals which then started growing in size and spreading on the Earth.

Primates — the group of mammals including *lemurs, monkeys* and, *apes* — 50 million years ago, had already split into *Platyrrhine*[43] monkeys in the New World (South America) and *Catarrhine* monkeys in the Old World (Africa)[44]. Thirty million years ago, the latter group in turn branched into Old World monkeys and great apes, which branched again around 15 million years ago into hominids, gorillas, chimps and orang-utans. Based on fossilized remains, found along the Middle Awash River in Ethiopia[45], the human family during the last 6 million years can be broadly divided into: *Ardipithecus, Australopithecus* and *Homo*.

Despite the separation which occurred 6 million years ago, the present genetic makeup of humans and chimps is still 99 percent identical. Data collected from fossilized ancestors do not provide a detailed reconstruction yet. It is clear however that hominids, chimpanzees and bonobos are descendants of Catarrhyne monkeys that lived in Africa.

The progressive cooling of the climate between 6 and 3 million years ago influenced life in general, probably being the cause of the separation between groups, some of which adapted to the conditions of the savannah and others like chimpanzees, which opted for the rainforest. Figure 4.1 shows the evolution of hominids.

Hominids were organized into small groups, but little is known about their life. The erect position (bipedalism) probably developed 3 to 4 million years ago as a result of local dry conditions. The discovery of fire and the Stone Age date back to around 2.6 million years ago when the first stone tools appeared.

Around 1.5 million years ago, hominids began to migrate in large waves towards Europe and Asia. From 2.5 to 1.5 million years ago the climate started varying with a general trend towards colder conditions and 4°C oscillations between warm and cold stages (Figure 5.3).

---

43 *Platyrrhine* is a descriptive term for monkeys with broad flat noses presently living in South America, while *catarrhines* are Old World (Africa and Europe) monkeys, ancestors of chimps, bonobos and hominids.

44 The separation of South America from Africa had already started 200 million years ago (Figure 3.2).

45 *The Evolutionary Road*, by Jamie Shreeve, National Geographic, July 2010.

FROM PREHOMINIDS TO HOMO SAPIENS

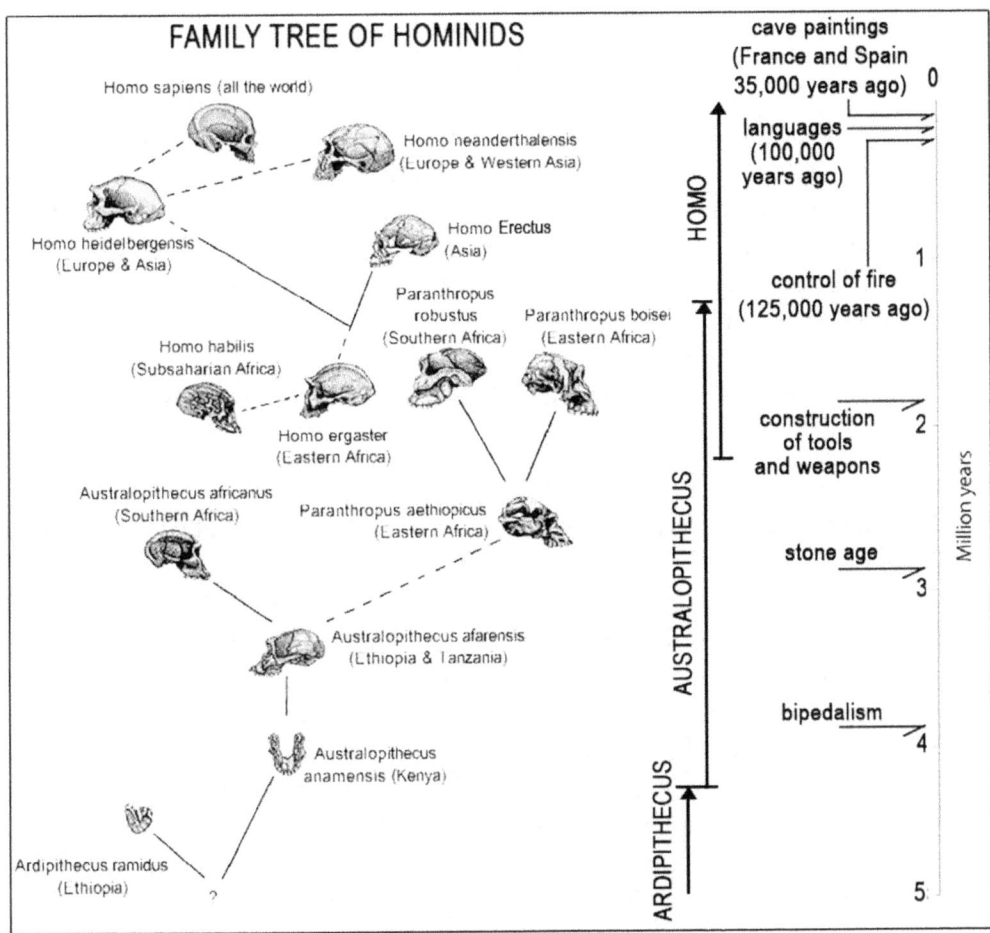

Figure 4.1 – The family tree of hominids (left) and some major events in human evolution (right).

The associated drought conditions and the scarce vegetation were at that time the major causes for herbivores to migrate towards grass-covered areas, followed soon after by carnivores. Scientists estimate that humans developed language between 100,000 and 50,000 years ago.

Hominids are basically divided into:

- Ardipithecus between 6 and 4.2 million years ago.

- Australopithecus from 4.5 to about 1.8 million years ago, probably a period of dry climate which influenced the diffusion in Africa.

- Homo during the last 2.2 million years. Homo Habilis, who lived from about 2.2 to 1.5 million years ago, is considered to have started the stone age, and the construction of utensils and prehistoric weapons. Homo Erectus reached Asia about 1.5 million years ago and his remains, found in China, date back to 1 million years ago. Homo Neanderthalensis, whose remains were found in Europe, the Middle East and Asia, lived from 400,000 to 40,000 years ago. *Homo sapiens* left Africa 100,000 years ago reaching Europe 50,000 years ago, when Homo Neanderthalensis was dying out.

The Wurm Ice age, Farming and Industrial Revolutions

During the Wurm glaciation – the coldest stage of which extended from 110,000 to 20,000 years ago — our ancestors arrived in Europe around 50,000 years ago when the coldest condition was in full development and the survival in the very harshest local conditions was unavoidable. The kilometre-thick Wurm ice cap (which covered the Northern hemisphere), began melting about 20,000 years ago making the sea levels rise about 110 m in the following 10,000 years worldwide. Disappearing glaciers and ice covers on the one side left behind huge quantities of glacial deposits named moraine[46] and numerous piedmont lakes in North America and along the Alpine-Himalayan range, on the other making the sea level rise on average 10 m every 1,000 years.

From about 20 to 10 ky ago the mean surface temperature went up about 10 °C and the rapid transition from a cold to a warm climate caused intense morphological changes, reshaping riverbeds, coastlines and mountain slopes.

Territories no longer covered by ice, were rapidly colonized by vegetation (forests and grasslands) and soon after by animals. Humanity's major development coincidentally occurred during the last 10,000 years of unprecedented stable climate conditions which lasted until half a century ago. Due to this unusual stability stage and the associated civilization process, the current risk of climate change is

---

46 **Moraine are chaotic accumulations of loose material deposited by the melting of glaciers.**

not yet felt by most people to be a real dramatic possibility, despite daily news about global warming and natural disasters.

Noah's flood described in the Genesis, was most probably associated with the last phenomena of deglaciation which persisted until around 7,500 years ago in the area of the Black Sea, as described by W. Ryan and W. Pitman[47].

The last 7 millennia were characterized by a rather stable climate stage, during which civilizations developed worldwide. The melting between 20,000 to 10,000 years ago, however, caused the sea levels to rise worldwide by about 110 m, which totally re-designed coastlines everywhere.

The stable climate during the last 10,000 years began changing around the 1800s when the industrial revolution started. As a consequence the consumption of coal began increasing rapidly and the mean surface temperature entered a rising phase, in turn driving the current global warming. The oldest settlements related to initial farming activities, dating back to 10,000 years ago, were found in the fertile zone between the rivers Tigris and Euphrates (Iraq).

The Neolithic agricultural revolution was a breakthrough in the historical human pathway, the beginning of a new interaction between man and nature and today it is considered the key to understand the current global environmental crisis. Agriculture is the first form of technically advanced activity, being based on a collective, although initially isolated farming experience of nomadic groups.

The Farming Revolution triggered a rapid knowledge-based process which affected the entire domain of human activities, including the understanding of seasonal cycles[48], the development of farming and irrigation techniques, written languages and communication.

Two fundamental aspects marked the change: the inherited struggle for survival of hunters-gatherers and a new vision of life based on human beings' ability to develop agriculture and create a more hospitable environment.

Around 7,000 years ago, the successful results achieved in agriculture led to the construction of organized cities in the Middle Eastern empires, later on followed by a significant cultural development in the Mediterranean region. A similar development coincidentally occurred in China, India, Central and South America, of course under different environmental and historical conditions.

The second and most recent change in human evolution was the industrial revolution, which started around the 1800s, based on a number of coincidences such as

---

47   *Noah's Flood. The new scientific discovery on the event that changed history*, 1998, by W. Ryan and W. Pitman.
48   Seasonal cycles refer to periods of favourable temperatures or rainfall, flooding, low water levels in rivers, seed cycles.

the explosive progress in science and technology, the invention of steam-powered and, later on, internal combustion engines which revolutionized the entire society.

The current change — that is the transition to the III Millennium — sounds more impressive, since it is abrupt, planetary in the size of human population, impacts and the depletion of basic resources.

The Migration of Homo Sapiens

Remains of *Homo sapiens* in Africa date back to 200,000 years ago, the migration started around 100,000 years ago and their spread over the planet took place during the last 50,000 years.

Figure 4.2 shows the migration from Africa towards the north-east, most probably as a consequence of the Riss-Wurm interglacial stage (or Eemian interglacial stage, 130,000 years ago, see Figure 5.6), when Africa's high temperatures were much like those of today and drought conditions were common.

Remains of *Homo sapiens* in Europe are dated back to 45,000 years ago, the time when the last Neanderthals are believed to have disappeared.

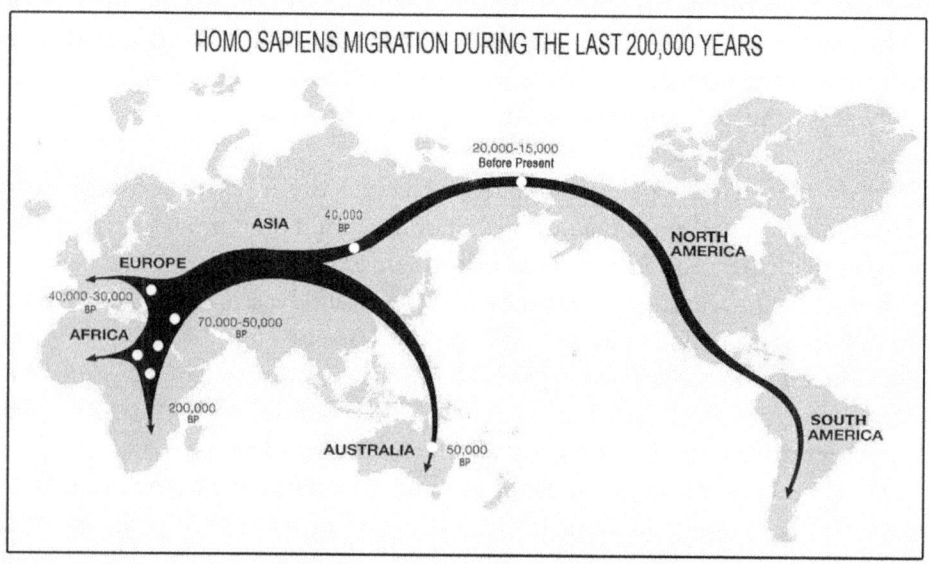

Figure 4.2 – Migratory waves of Homo sapiens over the last 200,000 years: (1) origin in East Africa (Ethiopia, Kenya, Tanzania), (2) crossing of the Red Sea, (3) diffusion in Asia and Australia, (4) arrival in Europe, (5) diffusion towards Siberia and (6) crossing of the Bering Strait and spreading into the Americas.

Scenes of hunting activities (cave paintings and clay modelling) by our ancestors — discovered in France (Chapelle aux Saints) and Spain (Altamira)[49] — date back 30 to 40 thousand years.

The Bering Strait was crossed some 20,000–15,000 years ago, when the Wurm icecaps had just begun to melt and while the sea level was still sufficiently low to preserve the land bridge of the Aleutian Archipelago between Siberia and Alaska.

Through the coordination of paleo-anthropologic data of fossil remains and the current distribution of DNA in the human population, the African origin of the human species has been confirmed. Genetically speaking, we humans can therefore consider ourselves to be one single family.

During the last part of the Wurm glaciation several local extinctions took place: 30,000 years ago, some great mammals disappeared, such as the musk ox and the sabre-toothed tigers, and 5,000 years ago the last mammoths became extinct. These large mammals, that had developed a thick insulating layer of fat over the previous 90,000 years of cold climate, could not adapt to the rapid and progressive rise of the temperature.

The first moderate production of greenhouse gases can be traced back to the Agricultural Revolution. As W.F. Ruddiman describes in his article on global warming[50], the emission of carbon dioxide in the atmosphere from agricultural practices coincides with the beginning of farming activities in the Mediterranean around 8 thousand years ago. Deforestation, burning of vegetation and rice cultivation in wetland fields and in more recent times the abundant use of fossil fuels contributed to the first significant greenhouse gas emissions during the past 2 centuries.

Since the industrial revolution (1800) the expansion of human activities was increasingly based on the use of steel, chemicals and coal and on important discoveries like the steam engine and later on the combustion engine. Environmental impacts became consistent between 1870 and 1900, when huge ships began transporting people and goods across the Atlantic, steam engine trains were developed in Europe and USA and modern industrialization began.

---

49 The primitive astonishing art expressions of our ancestors in Europe are dated as follows: (i) 30 ky ago the Musk-ox "flowing lines" with fingers in damp clay, Altamira (Spain), (ii) 25 ky ago Cueva del Castillo (Santander, Spain) with human hands over a bison and other animals, (iii) 20 to 10 ky ago clay reliefs of mating bison, Le Tuc d'Audoubert (France), (iv) 20 to 10 ky ago Cave Paintings of deer in Lescaux (France) and Altamira (Spain). Human intelligence and creativity was already advanced between 30 to 10 ky ago when our ancestors were confined to caves due to the severity of climate conditions. Source: Prehistoric and Primitive Man, 1966, The Hamlyn Publishing Group Limited.

50 Scientific American, May 2005

This fostered in turn the exponential growth of population, the productions and consumption of goods, the business as we know it today and the exponential rise of GDP (PPP). Human trends based on the speed of motion, population growth and world GDP (PPP) growth are represented in Figure 4.3 as indicators of the abrupt change.

The industrial revolution by replacing manual labour with machinery increased dramatically the human production capacity affecting in different ways every aspect of life. Major changes concerned agriculture, mining, manufacturing, culture and technological innovations, such as textiles production and the innumerable applications of steam powered engines.

Soon after the end of World War II environmental impacts, associated with the industrial development, were already felt in most industrialized urban areas and by the end of 1950s some premonitory local signs of environmental deterioration too.

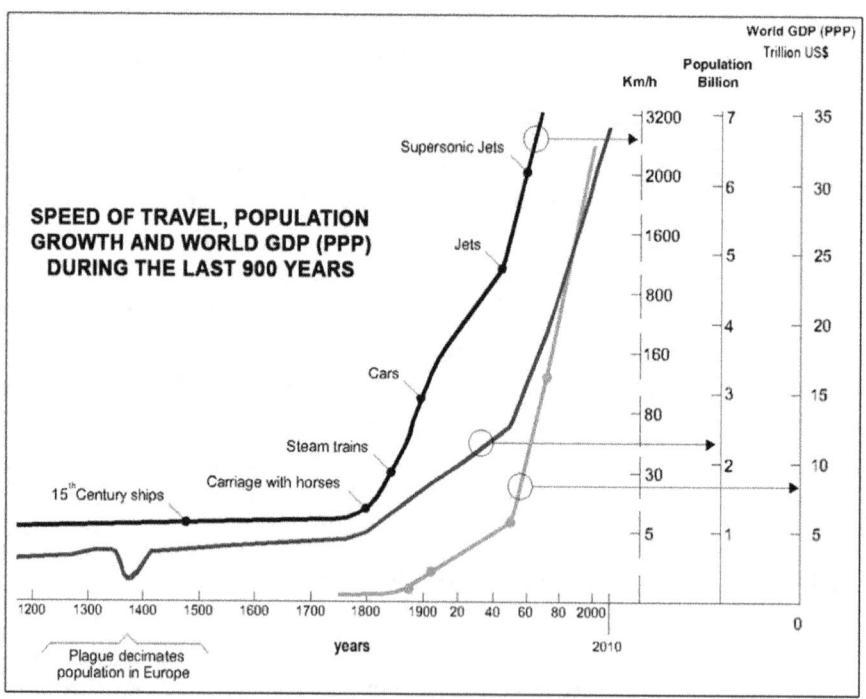

Figure 4.3 – Variations in the speed of communication, population and world GDP PPP[51]. The abrupt change began with the industrial revolution in 1800 and went on with an exponential trend. Data concerning speed of travel and population are derived from the 1971 book Quale Futuro? (Which future?) by Aurelio Peccei, Mondadori.

---

51   GDP PPP: Gross Domestic Product, Purchasing Power Parity

CHAPTER 5

Global Warming and Climate Change

Climate Change Factors

Introduction

The major driving factor behind climate is solar energy, which warms the Earth making it habitable to life by providing constant insolation[52]. Figure 5.1 shows the interacting components which today influece the climate on Earth. The interplay of extra-terrestrial and terrestrial factors, during the evolution of the Earth, determined the quantity of the solar energy reaching the planet, and

---

52 Insolation is the incident solar radiation that reaches the top of the atmosphere, expressed in W/m^2.

the absorption by the atmosphere and the surface of the Earth, affecting the mean surface temperature over time.

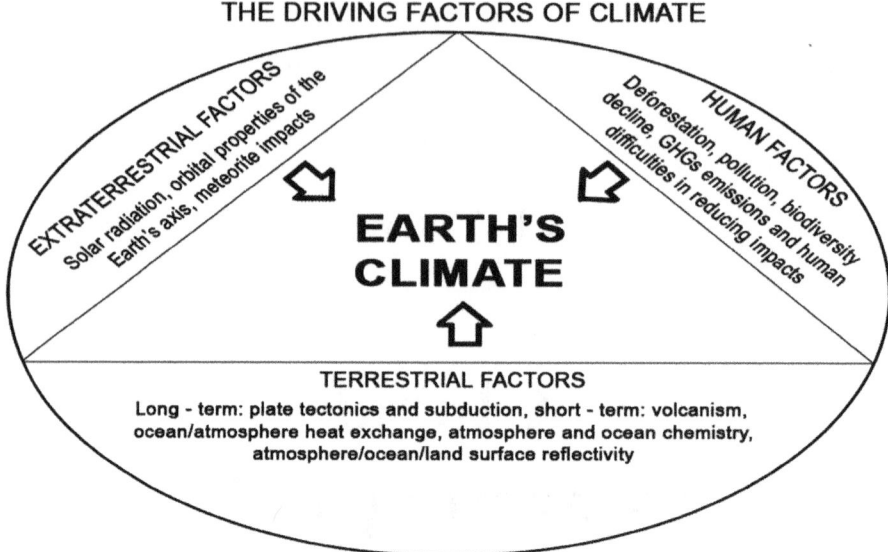

Figure 5.1 – Factors of climate change.

The interdependence of factors has been heavily perturbed during the last two centuries, and more deeply in the last 60 years, by the growth of human activities with the potential to alter the stability of climate and the environment. Humanity — the emerging force in the geological biological domains — started accepting the responsibility of global impacts, pollution and the disruption of natural cycles during the last three decades.

Extraterrestrial factors

Extraterrestrial factors include:

- solar radiation, which reaches the Earth in a quantity compatible with life, amounts to 342 W/m^2, out of which 240 W/m^2 are absorbed by the planet and 102 W/m^2 are reflected into the space. The daily rotation of the Earth exposes half of the planet's surface to solar radiation every 12 hours, preventing the overheating and overcooling of the biosphere;

- orbital cycles, which influence the distribution and the amount of solar energy at the top of the atmosphere and alter the energy absorption by the Earth. The cycles are: (i) the eccentricity of the Earth's orbit which ranges from a nearly circular to an elliptical shape with major periods of 413 ky[53], 125 ky and 95 ky, (ii) the obliquity of the axis varying between 22° and 24.5° in 41 ky and, (iii) the precession in the rotation of the Earth's axis with a cycle of 19–24 ky.

- meteorite impacts, which drove huge destruction in the past and triggered climate changes of variable intensity. The two largest collisions — which caused a global catastrophe, an abrupt climate alteration and a mass extinctions — occurred at the end of the Permian and the Cretaceous, respectively 240 and 65 My ago.

The eccentricity of the Earth — with a periodic recurrence of about 102 ky (in the range of 95–125 ky) and a longer cycle of 413 ky — determines the variation of the Sun-Earth distance.

In about 100,000 years the distance varies from 147 million km, when the Earth is closest to the Sun, the orbit is elliptical (0.058) and warm conditions prevail, to 152 million km when the orbit is nearly circular (0.005) and the atmosphere is cooler. The current eccentricity is 0.017, which implies a relatively circular orbit.

The strict correlation between eccentricity variation and Ice Age cycles during the last 650 ky is illustrated in Figure 5.2. While eccentricity (top) shows a harmonic shape, the Ice Age graph (bottom) depicts a more fragmented trend due to the interactions between eccentricity, the obliquity of the Earth's axis, the precession of equinoxes and a variety of terrestrial phenomena. Human impacts, associated with the 10,000 years old farming revolution, and the more invasive anthropogenic activities and effects due to the industrial revolution during the last two centuries, added up to the traditional interplay of climate change factors.
According to Milankovitch, when the orbit approaches the maximum elliptical shape, the Earth is closer to the Sun, a greater quantity of energy is absorbed, seasonal differences are bigger and the surface temperature is globally higher. When the orbit is circular, the distance between Earth and Sun is greater, seasonal variations are minimal and climate is globally cooler.

---

53 Time in the geological domain can be expressed in full as 500,000 years or as 500 ky (thousand years) or 0.5 My (million years), followed by BP for Before Present (1950).

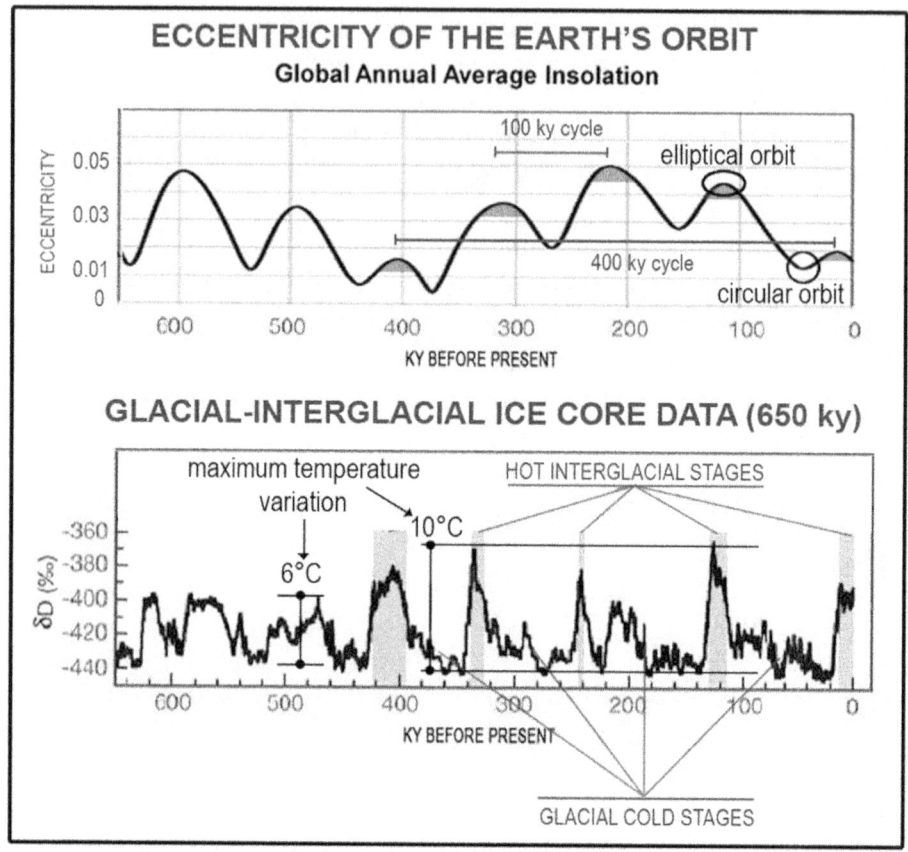

Figure 5.2 – (top) Eccentricity variations from a circular to an elliptical orbit. (bottom) δD (deuterium) variation, as an indicator of temperature changes and climate variations. The graph shows Ice Age cycles during the last 650,000 years. Each cycle of about 100,000 years comprises a cold stage of 80-90 ky and a warm stage of 20-10 ky. The temperature oscillated about 6°C between 650 and 450 ky, and around 10°C during the last 450 ky. Adapted from: IPCC Climate Change Report, 2007 AR4.

The eccentricity of the orbit is primarily influenced by the interaction of Jupiter and Saturn gravity fields. During the last 450,000 years, the alternating occurrence of elliptical and circular shapes generated the sequence of glacial cold and warm stages, during which, respectively, ice covers expanded lowering the sea level and retreated due to ice melting, in turn cause of global sea level rise. By interacting with the eccentricity, the obliquity of the Earth's axis and the precession cause insolation to vary and play a fundamental role in modulating surface temperatures and climate variations. Changes in the obliquity of the axis significantly affect seasonal contrasts, while the precession of equinoxes determines the position of the Earth's poles relative to the sun .

Terrestrial factors

The terrestrial driving factors of climate change are:

- ocean variability, which is the main driver of ocean water circulation, marine biological activity, seawater carbonate chemistry, sea-air heat exchange and seawater and atmosphere chemistry

- CO_2 abundance, surface and deep-water temperature variations drive changes in atmospheric CO_2; surface biochemical processes affecting the carbon cycle and sedimentary carbonate deposits on the ocean floor which, by subtracting huge quantities of calcium carbonate, modify the global carbon balance

- the expansion of surface vegetation, which increases the absorption of solar radiation during warm periods, and snow and ice covers during glacial stages, which reduce absorption by increasing the reflection of light. Oceans and land reflectivity, is named albedo[54]

- volcanism as a source of heat, dust, water vapour and other gaseous emissions during eruptions. Most of the contribution is from lava flows along underwater fractures in the Mid Atlantic and East Pacific Ocean ridges (see Chapter 3). Continental drift and plate tectonics are responsible for the long-term build-up of mountain chains, which also influence the distribution of climatic zones on the Earth

- water circulation in the atmosphere and the associated heat transport

- the expansion of life and the formation of coal, oil and gas deposits, which remained underground during tens of million years, subtracting from the biosphere huge quantities of carbon-rich biological material

- the atmosphere-ocean interaction, bio-geochemical cycles, the extent of cloudiness and the abundance of atmospheric dust.

---

54   The albedo effect is the reflectivity potential of ice covers and other materials on Earth.

Human factors

The current driving human factors of climate change are:

- the huge and abrupt emissions of CO_2 and other GHGs, associated with human activities during the past two centuries, which have had a relevant impact on climate. Deforestation, pollution, the human invasion of nearly every habitat on Earth and the growth of population made impacts to increase in intensity and accelerate the greenhouse effect

- the difficulty for human beings to change the current economic and social development trend; unlike other living organisms, humanity can survive under any local environmental condition, provided that resources are available.

Research Methods and Effects of the Climate Change

Two integrated research methods are followed in the climate domain:

- the paleoclimate approach based on proxy data[55] from ice cores (from Arctic, Antarctic and Greenland projects), studies of ocean sediments, corals and tree rings. Temperature variations are derived from studies on δD (deuterium) and ^{18}O, glaciers and their moraine deposits[56];

- computer models, which use quantitative methods to fill the gap between local and global scales, simulating interactions among the atmosphere, oceans and land surface, and providing future scenarios.

The description of climate change is mainly centred on paleoclimate data from ice and sediment cores. The geological signs left by climate on surface rocks provide useful information on the evolution of the environment, climate variations, plant and animal life.

The most promising studies are based on the analysis of dust, air bubbles, the quality of pollen and isotopes (^{13}C, ^{18}O, Deuterium and Beryllium) from ice

---

55   Proxy data are derived from different indicators (ice cores, corals, tree rings, boreholes) and are used as a crossing verification tool in climate change studies.
56   A moraine is an accumulation of materials abandoned by glaciers during the melting stage.

and sediment cores, see floor mud, and fossil rich layers, from Greenland and the Vostok Station in Antarctica and other places. Climate change can worsen or improve atmospheric conditions, drive impacts on life and the environment, and carry the potential to disrupt current human activities. If global warming continues in the next years reaching 2°C above preindustrial era, then either the global temperature will keep rising to 3–4°C, or it will decrease abruptly causing a reglaciation process, in both cases with dramatic effects on our society. Today, a World 2°C warmer is considered a very critical condition. The UN Climate Summit held in Durban (South Africa) in November 29, 2011, has not been able to reach an agreement on fund rising for developing countries to reduce CO_2 emissions. The US, Saudi Arabia, the EU and other rich countries disagreed on the possibility to rise $ 100 billion by the end of this decade, as recommended by the Green Climate Fund. The Kyoto Protocol expires in 2012, therefore, once more vulnerable developing nations will pay the highest price. Procrastinating important global decisions inevitably leads to catastrophic events, growing worldwide in number and intensity. During the Phanerozoic, which lasted 542 My (Figure 3.4), the Earth's climate[57] underwent a series of changes, warm conditions broadly covering two thirds of that time and cold conditions one third of it. Three fourths of the 65 My long Cenozoic Era[58] were characterized by a stepped climate cooling (Figure 5.3). During the latest five million years — when hominids evolved (Figure 4.1) — climate changes heavily affected the environment:

- the distribution of ecosystems (plants and animals) was constrained by changes which reduced or improved the stability of the environment and life, depending on local conditions, and on the availability of nutrients and water. Due to climate changes life adapted, migrated or went extinct;

- the Earth's physical features were sculptured by ice age cycles, sea level changes, erosion and sedimentation, the bio-geochemical weathering of rocks and the opening (or closing) of terrestrial intercontinental bridges.

---

57 Climate is defined as a complex system influenced by temperature, atmospheric pressure and humidity, wind, rainfall and other meteorological aspects over several decades in a region.

58 The Cenozoic, which means "recent life", is the last geological era (from 65 million years ago to the present) and includes the Tertiary and the Quaternary periods. The Tertiary is divided into Paleocene, Eocene, Oligocene, Miocene and Pliocene, the Quaternary into Pleistocene and Holocene.

During the last ice age (Wurm), the Aleutian Islands, as a consequence of low sea level, bridged Siberia to Alaska.

The current global warming is affecting the biosphere at a growing speed, causing massive threats to our complex society. Migrations of invasive species are alarming and their effects on biodiversity too, since local conditions are modified faster than adaptation by living organisms does.

Climate Change

The cenozoic era

Evidence of climate change throughout the geological history is unquestionable and signs of past and recent variations are imprinted in the Earth's sedimentary rocks. The Mesozoic Era (Triassic, Jurassic and Cretaceous) which lasted from 250 to 65 Mya, encompasses the assembling of continents into the Pangea (Figure 3.2) 225 Mya, the rise of angiosperms (flowering vegetation), the spread of dinosaurs worldwide, a global climate predominantly hot and arid and low sea levels. Between 225 and 200 Mya, the breakup of Pangea and the formation of the Thetis Sea fostered the separation of the landmass into two major continents: Gondwanaland in the South and Laurasia in the northern hemisphere. Global climate during Mesozoic was dominated by hothouse conditions and a rising temperature which peaked in the Early Cenozoic, that is in the PETM 56 Mya and the 52 Mya critical point of inversion towards a cooler climate. Figure 5.3 provides an overview of global climate changes during the Cenozoic Era.

The history of Cenozoic variations of carbon dioxide and temperature helps in understanding the connection between climate drivers, radiative forcing and climate changes, therefore providing a useful information for the trend of the current climate change.

Graph (a) shows climate conditions during the Cenozoic Era: (i) extreme warming between 65 and 46 Mya, an iceless period of global tropical climate marked by the Paleocene-Eocene Thermal Maximum or PETM 56 Mya, and the 52 Mya temperature inversion point, (ii) the long term cooling period, during which 37 Mya and 7 Mya the polar ice sheets formed. The growth of the Antarctic Ice Sheet (AIS) was halted between 27 and 14 Mya by a small temperature rise, which partially melted it. The AIS was again thickened by reglaciation during the last 13 million years.

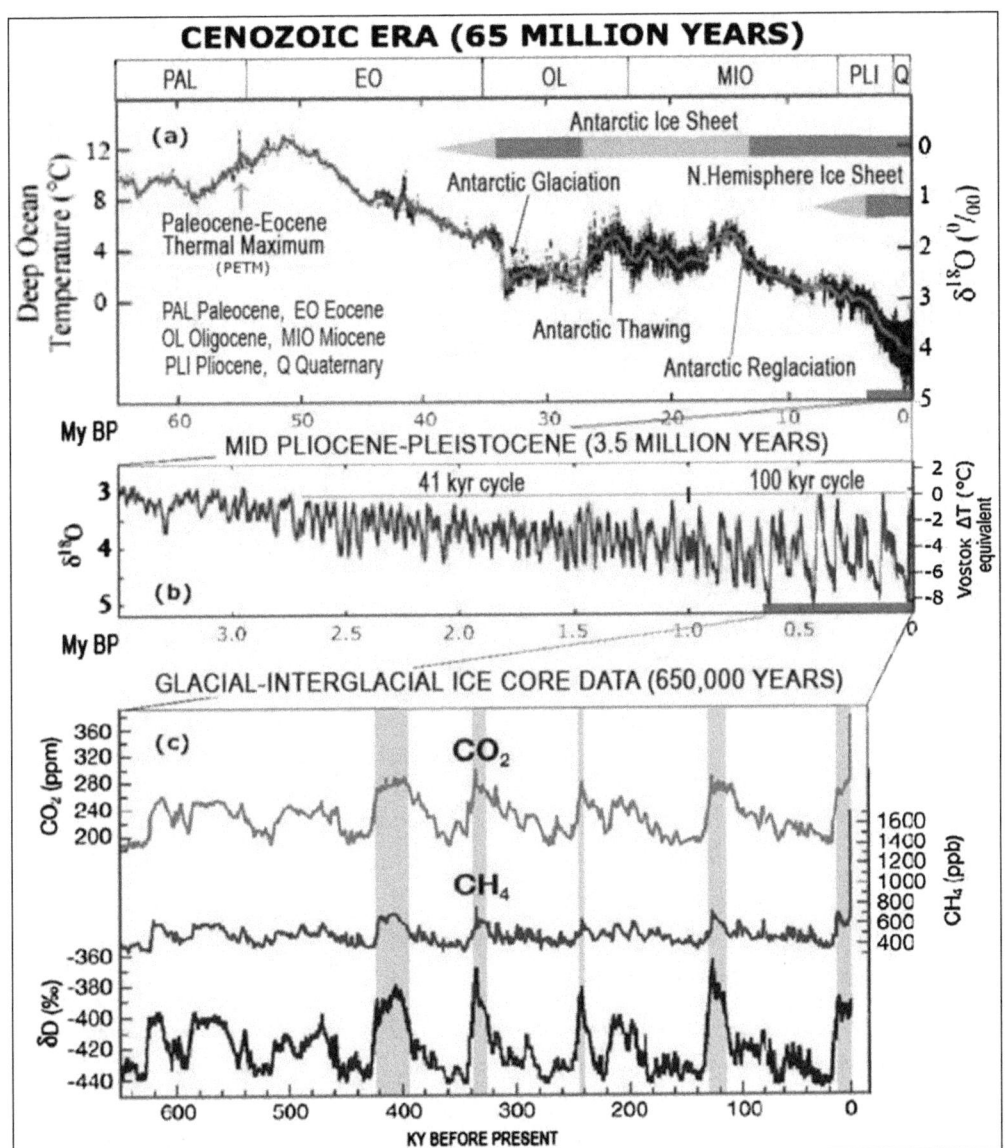

Figure 5.3 – Overview of global climate change during the Cenozoic Era. Graphs (a) and (b) show temperature variations based on ^{18}O isotope composition from fossilized shells of benthic foraminifera. Graph (a) also highlights the formation of Antarctic and Arctic ice sheets (dark ice bars correspond to deep freeze conditions, compared to the light gray bar indicative of the Antarctic thawing, between 27 and 14 Mya). Graph (c) shows variations of CO_2 and CH_4 atmospheric fluctuations and δD (deuterium), a proxy for local temperature from Vostok (Antarctica) ice cores, during the last 650 ky. Sources: (a) Zachos et al. 2001), (b) Lisiecki L.E. and M.E. Raymo: A Plio-Pleistocene stack of 57 globally distributed benthic d 18O records. Paleoceanography 20: 103, doi: 01029/004PA001071, (c) IPCC Climate Change Report 2007, AR4.

Graph (b) shows temperatures (right scale) during the last 3.5 million years, based on ^{18}O variations in Vostok ice cores (left scale). Oscillations were mainly influenced by orbital factors: the 41 ky cycle due to the obliquity of the Earth's axis and a 100 ky cycle due to the eccentricity of the orbit[59] in the last three million years.

Graph (c) illustrates variations of CO_2, CH_4 and δD^{60} (Deuterium) during last 650 ky (ice ages), grey bands highlighting the correlation among the three indicators. Figure 5.4 shows in detail abrupt changes caused by human-related emissions of CO_2, CH_4, N_2O and the associated temperature rise during the last 200 years.

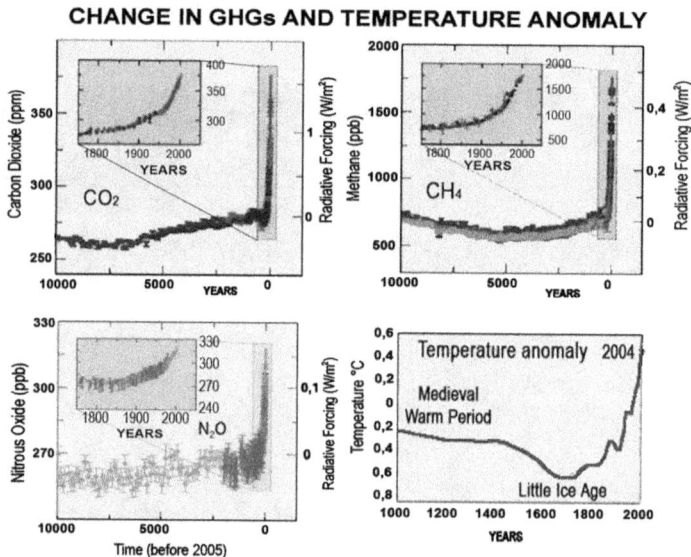

Figure 5.4 – CO_2, CH_4 and N_2O variations during the last 10 ky and the last two centuries, and a 1°C temperature rise since the 1750s. Adapted from IPCC, Climate Change Report, AR4 2007.

The present glacial condition of the Earth — characterized by a cyclic sequence of ice ages during the last one million years — is the result of the long-term cooling trend which initiated in the Early Cenozoic. The abrupt rise in temperature during

---

59 Obliquity and eccentricity changes are attributed to the interaction of Jupiter and Saturn gravitational fields.

60 δD Deuterium concentration differences. Deuterium concentration in Vostok Antarctica ice cores has been used to derive local temperatures relative to 1990 (Petit et al., 1999).

the last 2 centuries, after 10 ky of relative climate stability, is therefore, considered exceptional.

Recent human activities and the related GHGs emissions, are today held responsible for the temperature anomaly and global warming. The power of human forcing highlights the sensitivity of climate and the possibility that further temperature rise may trigger dramatic effects on our overpopulated planet. The concentration of CO_2 over 1000 ppm, was essential in driving the very hot climate in the Early Cenozoic. When CO_2 began decreasing around 52 Mya, the lowering concentration caused a long-term cooling, which dominated almost entirely the Cenozoic, CO_2 reaching a stable level (180–280 ppm) during the last one million years. Huge CO_2 emissions in the Early Cenozoic are mainly attributed to tectonics:

- the long-term drift of continents (Figure 3.2), towards their current positions, included India's collision against Asia (critical between 57 and 49 Mya), which caused the uplifting of the Himalaya Range and the Tibetan Plateau. The crush was the consequence of a long-term migration (20 cm/year) of India, which compressed the pre-existing Tethys Ocean (Figure 3.2), squeezing sediments and the organic matter accumulated on the ocean floor. Sediments and the Tethys oceanic crust, in the most critical phase of the crush, underwent subduction below the Asian plate. The related metamorphism produced a huge degassing of CO_2 which was abruptly released 56 Mya (PETM). Massive emissions of magma took place during the most critical period;

- the widening of the Atlantic Ocean, the separation of Greenland from the British Islands and the northern extension of the Mid Atlantic Ridge (due to the divergence of plates), which enhanced the rise of magma from the upper mantle (the model is similar to Figure 3.3 in the Pacific).

In sum driving factors shaping the climate operate at different timescales and intensities, altering the global energy balance and causing temperature variations and climate change[61]. The powerful force of tectonics in the Early Cenozoic and

---

61 It is worth recalling the huge climate change in the Mediterranean Region between 7 and 5 Mya. As a consequence of the formation of Gibraltar Isthmus, due to the collision between Africa and Europe, the Mediterranean became a closed basin and its coastal areas underwent desertification. W. Rayan and W. Pitman's 1998 book "*Noah's Flood*", describe surprising geological discoveries, as the 200 m decrease in the Mediterranean's sea level and the formation of thick salt layers in the sea floor. Five million years ago the Isthmus opened up again and for 50,000 years water from the Atlantic filled up the basin again.

the associated huge emission of carbon dioxide caused the global warming and an iceless tropical climate condition in polar regions. Fossils, recovered in Alaska (crocodile-like reptiles) and the Central Siberia, provide a clear evidence of warm global conditions and depict a nearly iceless planet. Soon after the end of PETM (which is a spike in geological terms), temperature rose again until 52 Mya, when the tectonic forcing of India against Asia started attenuating, CO_2 emissions began decreasing and conditions for a temperature inversion from hot to cold prevailed. The current trend of climate on Earth is considered similar (although much faster) to the sudden change of temperature and the greenhouse forcing during the PETM. In fact GHGs emissions during the last decades are growing at a much higher speed than in the Early Cenozoic.

The paleocene-eocene thermal maximum or PETM

During the Early Cenozoic (PETM), temperature rose 6°C causing the extinction of benthic foraminifera and some terrestrial mammals, CO_2 reached 1500 ppm, and the excess of carbon was absorbed by forests and oceans during over 100,000 years.
According to J. Hansen and M. Sato[62], CO_2 level increased during the PETM at least to 1000 ppm, declining to 170 ppm during recent Ice Ages, when the geological forcing associated with the crush of India against Asia was no longer critical. The climate forcing computed by Hansen and Sato for the above CO_2 range exceeded by 10 W/m² the forcing during the Ice Ages, CO_2 therefore resulting the predominant climate driver in the Early Cenozoic. The quantity of CO_2 released today into the atmosphere, due to the consumption of fossil fuels, is 30 billion metric tons per year (MTY), compared to 2 billion MTY during the PETM, being therefore 15 times greater and driving temperature up at an unprecedented speed. Scientists expect that CO_2 will rise further and faster during next decades, unless humanity makes a drastic transition to renewables. Cenozoic climate changes carry serious implications for our complex society:

- the presence of the polar ice sheets during the last 37 and 7 million years on Earth caused a greater reflectivity (albedo effect) of solar radiation, and the long-term decreasing of mean global temperatures worldwide. The current anthropogenic melting of the polar Ice sheets, instead, by

---

62 J. Hansen and M. Sato: Paleoclimate implications from Human — Made Climate Change.

reducing reflectivity and enhancing the absorption of solar energy, leads faster towards a warmer planet and a hothouse condition;

- the long-term Cenozoic cooling involved major and minor climate change drivers, their interactions being complex and not yet fully understood. Human interference and impacts comparatively appear concentrated in the short time scale of few decades, significantly more predominant in dimensions and unstoppable in the short term.

The risks we are facing today are:

- the rapid rise of CO_2 (due to the sum of CO_2 from fossil fuels and emissions from the melting of tundra and permafrost), has already increased the greenhouse effect driving global warming. Climate change is advancing so rapidly that the reversibility limit might have been trespassed already

- sea level rise, hurricanes, catastrophic floods and other impacts threaten coastal areas worldwide; the non-linear melting of polar ice sheets can reach a critical condition

- ocean acidification, coral bleaching[63] and the ecosystem's decline

- food scarcity due to expanding drought conditions in arid regions

- insect infestation and diseases.spreading worldwide.

The PETM was first identified through the sudden change in the ratios of isotopes ^{12}C and ^{13}C in sediment core samples from the ocean floor near Antarctica. More recently, the transition was also visibly identified in core samples through the presence of white shells of foraminifera which, as a consequence of the PETM and the sudden rise of CO_2, abruptly turned into a light red colour, indicative of dissolved shells, due to ocean acidification[64]. Compared to the

---

63 Coral bleaching includes water temperature variations, increased solar irradiance and changes in water chemistry (acidification). A massive coral bleaching has hit the Great Barrier of Corals in Australia.

64 The change was identified in sediment cores holding the 56 Mya PETM event and the transition from plankton carbonatic shells to the absence of these shells due to the sea water acidification. National Geographic, October 2011, *World Without Ice*, by R. Kunzig.

temperature rise associated with the PETM, the current global warming is faster, carries the possibility of tropics to extend towards sub-polar regions and generates in the short term a variety of impacts. According to the palaeontologist D. M. Raup (*Extinction. Bad Genes or Bad Luck?*, 1991) the absence of coral barriers always followed past mass extinctions.

The 2°C overshooting and temperature inversion during ice ages

Carbon dioxide trends, temperature changes and the temperature inversion mechanism during the last 540 ky are shown in Figure 5.5.
Graphs (a) and (b) show similar shapes of CO_2 and temperature changes during the last 450 ky and the current exponential growth of CO_2 concentrations in the last 200 years.

Graph (b) illustrates in detail the *average temperature variation of about 10°C* during the last five temperature inversion stages (shown in the ellipses) and the *2°C overshooting temperature window*, that is the range in the middle of which humanity is currently trapped. The mean temperature is so close to the upper limit that the possibility of reaching safely 2°C overshooting and 450 ppm of CO_2 is simply shocking.

Graph (c) shows the bell-shaped stack from the overlapping of the past interglacial (ellipses in graph b), with coinciding temperature peaks:

- point (A) represents the temperature inversion point at the end of a glacial maximum stage, and the beginning of a progressively warmer trend (driver of ice melting), probably associated with the transition from a circular to an elliptical orbit of the Earth (B). The nearest (A) inversion occurred 20 ky ago ending 10 ky ago with a sea level rise of 110 m

- point (B) highlights temperature peaks 2–3°C above the pre-industrial era and the inversion from hot to cold conditions (and the beginning of a reglaciation), probably triggered by the shutdown of the Thermohaline Circulation[65] in the Northern Atlantic and a growing albedo effect

---

65   See Figure 5.10.

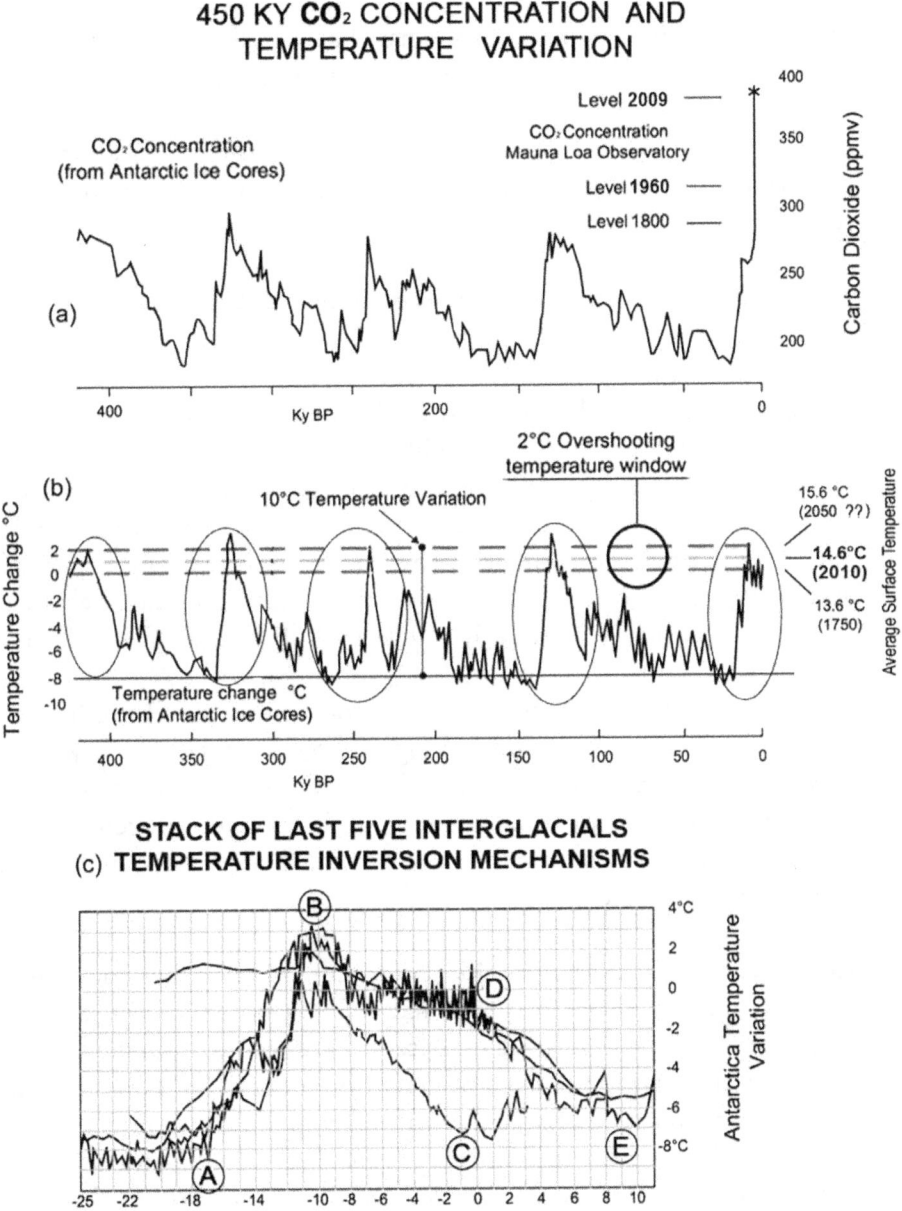

Figure 5.5 – *(a) CO_2 variations from 180 to 280 ppm during the last 450 ky and the abrupt rise during the last 2 centuries from 280 to 390 ppm (2009), (b) mean temperature variations during the last 450 ky and the 2°C overshooting window, (c) the stack of last five interglacials in the ellipses in graph (b). Sources: (a) and (b): variations in CO_2 and temperature from Vostok ice cores (NOAA), (c) adapted from: L.D. Roper: The Current Major Interglacial*

- point (C) represents the end of a reglaciation process after which deep freeze conditions continued in the past half a million years until the next deglaciation stage started, about 90 ky later, as in point A

- the segment (B)–(D)–(E) shows a long-term transition from warm to cold that could take place if humanity (at the present near D) would halt global warming.

Scenarios that can be envisaged on the basis of climate changes in the past ice ages are:

- if temperature – now 1°C above preindustrial time — keeps rising at the current trend (i) the atmosphere will turn into a "hothouse", life in general and the human society being affected by huge disasters (hurricanes, floods and wildfires), growing in quantity and intensity. Large emissions of methane into the atmosphere are likely to occur from the melting of permafrost and/or methane ice deposits on the sea floor; (ii) sea level is likely to reach within this century 1 m sea above 1900 level. (Business as Usual Scenario)

- on the basis of past ice ages, also a temperature inversion from hot to cold could take place as in the model in Figure 5.5 (c). This possibility, however, is considered unrealistic, since based on prevailing orbital factors, which seems not to be the case today: due to the abrupt anthropogenic growth of CO_2, human impact is prevailing as a climate change factor and a powerful driver of temperature rise

- the third scenario is the one in which humanity reaches soon an international agreement to drastically halt GHGs emission and stabilize the climate at the current level or so. After a few decades of further but slower warming, due to permanence of greenhouse gasses in the atmosphere, a stable condition of CO_2 and the mean temperature should be reached. This would in turn help humanity to proceed faster towards a sustainable carbon-free society. In practice CO_2 concentration in this scenario is expected to stabilize around 400 ppm or so and a temperature slightly over 1°C compared to the preindustrial era. This could significantly help the long pathway of humanity towards a sustainable development.

The sea level rise, in this optimistic scenario, would still increase but at a slower rate, probably reaching 1 m above 1900 level in the next century, depending on the speed of change of our society towards a carbon-free condition.

The last ice age

Events during the Wurm Ice Age[66], the arrival of *Homo sapiens* in Europe 40 Ky ago (the coldest phase), and the sequence of temperature spikes from 45 to 10 ky ago are shown in Figure 5.6. During the alternating sequence of spikes, our ancestors survived in a tight sequence of climate shifts known as the Dansgaard-Oeschger events, the temperature oscillation being on average about 6°C, each spike-arm lasting around 1470 years.

Spikes were either related to the instability of the ice bodies, or to ocean heat transport associated with the Thermohaline Circulation (see Fig. 5.10) or a combination of factors. The climate conditions during the last 10,000 years (when the agricultural revolution developed) can be considered as a "special coincidence", probably the result of the balance between a slow natural cooling trend and the human related global warming. Figure 5.6 (top) shows in detail climate changes in the last 150 ky, the migration of *Homo Sapiens* and the arrival in Europe, the deglaciation which started around 20–18 ky ago and went on until the beginning of the farming revolution.

Data concerning the Wurm Ice Age are derived from ice core samples from the Vostok Station in Antarctica (top), and ice cores from Greenland (bottom). Differences in temperature changes in the two hemispheres are significant; spikes related to the instability of the Greenland ice sheet represent more extreme conditions compared to variations from Vostok ice cores (Antarctica).

The bell-shaped mechanism of temperature variations between 139 and 108 ky ago (top) — the Eemian Interglacial — depict the model of temperature inversion that has characterized climate changes during the last 450 ky.

---

66 Ice core data come from the Russian Vostok Antarctic Research Station.

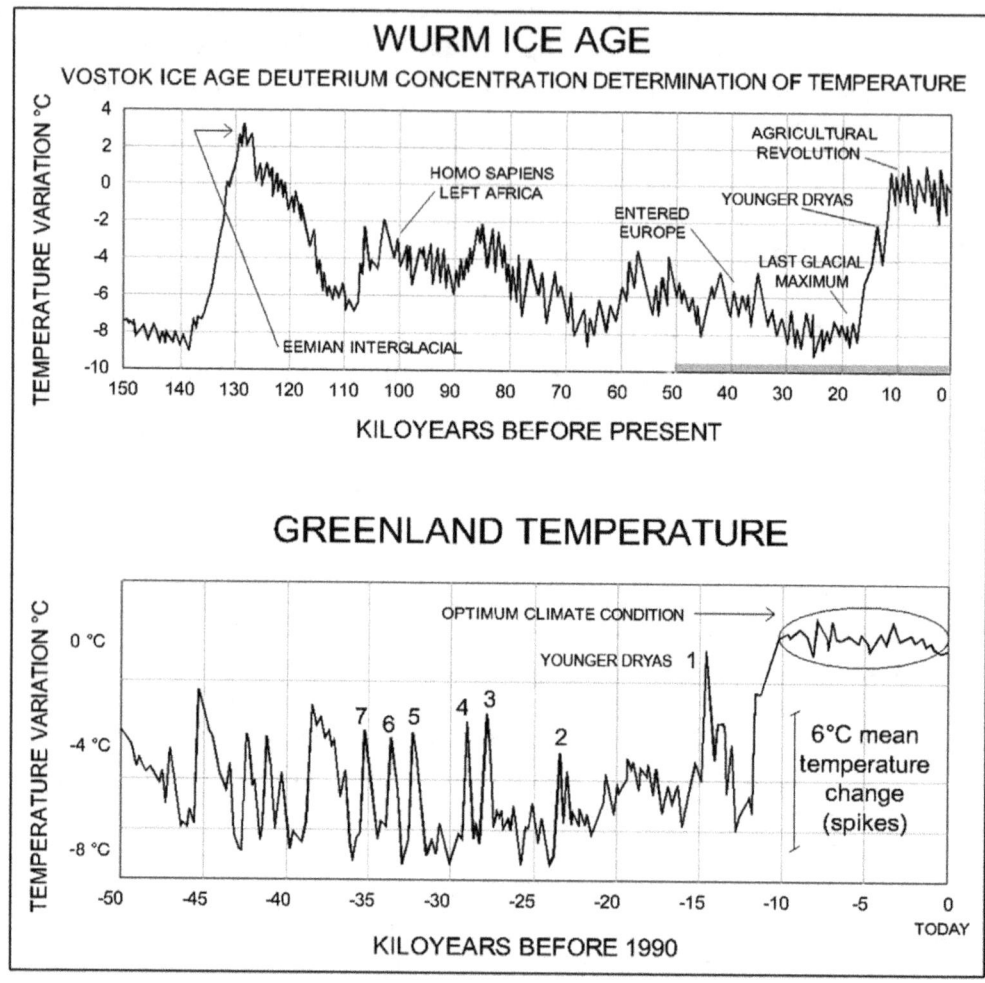

Figure 5.6 – (top) Wurm Ice Age, from Vostok Station data (Antarctica), and the migration of Homo sapiens from Africa. (bottom) the sequence of Dansgaard-Oeschger temperature spikes due to the instability of ice bodies, from Greenland cores. The coldest stage of Wurm developed between 110 and 20 ky ago when our ancestors had already reached Europe. Source: D. Roper The Current Major Interglacial

The end of the Wurm glaciation stage — marked in Figure 5.6 (top) as the Last Glacial Maximum (LGM) — was most probably caused by the transition from the Earth's nearly circular orbit (at which the Earth-Sun distance reaches the maximum), to a growingly elliptical orbit.

Relevant events which occurred during the Wurm are:

- the Eemian temperature inversion at 129 ky ago (from hot to cold), when the mean global temperature reached 3°C above our pre-industrial level, is the nearest and the best known interglacial model to which the current climate condition can be compared. If today's mean temperature keeps increasing, rapid ice melting, sea level rise and other climate related effects can take place, as they did in the Eemian

- the global see-level rise ranged between 4 m and 6 m, a condition which did not affect our ancestors in an underpopulated planet

- since their arrival in Europe 40 ky ago, our ancestors experienced dramatic difficulties due to extreme cold conditions and the sequence of temperature spikes. The Younger Dryas, a 1,400-year cold period, abruptly halted the deglaciation process in the Northern Hemisphere 14 ky ago, suddenly triggering a deep freeze

- the last glacial maximum (LGM) was reached around 18 ky ago, when the nearest event of temperature inversion from cold to warm occurred, ice sheets melted and in about 10 ky and sea levels rose worldwide 110 m. The temperature increase lasted until 10 ky ago, remaining nearly stable during the Farming Revolution. A new rapid temperature increase started 200 years ago as a consequence of the Industrial Revolution and the growing emissions of GHGs.

Evidence shows that paleoclimate and geological data from the Eemian, and, more specifically, from the last 20 ky, provide useful information on the huge impacts inevitably associated with the current climate change: ice melting, sea level rise, impacts on ecosystems and the people (nearly 1 billion) living today along coastal areas. The history of the ice ages highlights with undisputable evidence the sensitivity level of the climate system to driving factors and provides, at the same time, an indication of risks and mitigation measures to be adopted. The overall picture of the climate change, as it is inferred from the innumerable signs from the Earth[67] (morphological changes associated with

---

67 National Geographic, September 2004, Global Warning – Bulletins From a Warmer World.

ice melting, ecological signs affecting fauna and flora and time signs), implies that:

- anthropogenic activities are responsible for the current fast rise of the mean temperature and the global warming

- humanity is too slow in gaining awareness, changing its industrial production organization and the lifestyle, and getting involved in a low-carbon economy

- future generations will inevitably confront with a devastating climate.

The Atmosphere's Energy Balance and Global Radiative Forcing

Energy flow and the greenhouse effect

To explain the role of humanity as a driver of climate change and the radiative forcing trend[68], an overview of the global energy balance and budget is provided. Figure 5.7 shows the Earth's energy balance diagram. Out of the 100% incoming solar radiation 31% of that energy is reflected (8% by air, 17% by clouds, 6% by the Earth's surface), 69% is absorbed (19% by vapour, dust, ozone, 4% by clouds, 46% by the Earth's surface) and re-emitted towards outer space (9% + 40% + 20%) as infrared radiation during the 24 hours.

The greenhouse mechanism that controls the Biosphere's energy balance is tuned in such a way as to be compatible with life on Earth. The Biosphere is sufficiently warm and hospitable to living organisms thanks to the natural greenhouse effect[69], in the absence of which the mean temperature on Earth would have been constantly around −18°C, and the planet a cold, probably lifeless, place.

---

68 Radiative forcing is the net irradiance of solar energy, which is in turn the difference between incoming and outgoing radiation, measured in W per m^2.

69 The natural greenhouse effect is the result of to the contribution of four gases: water vapour 36–70%, Carbon Dioxide 9–36%, Methane 4–9% and Ozone 3–7%.

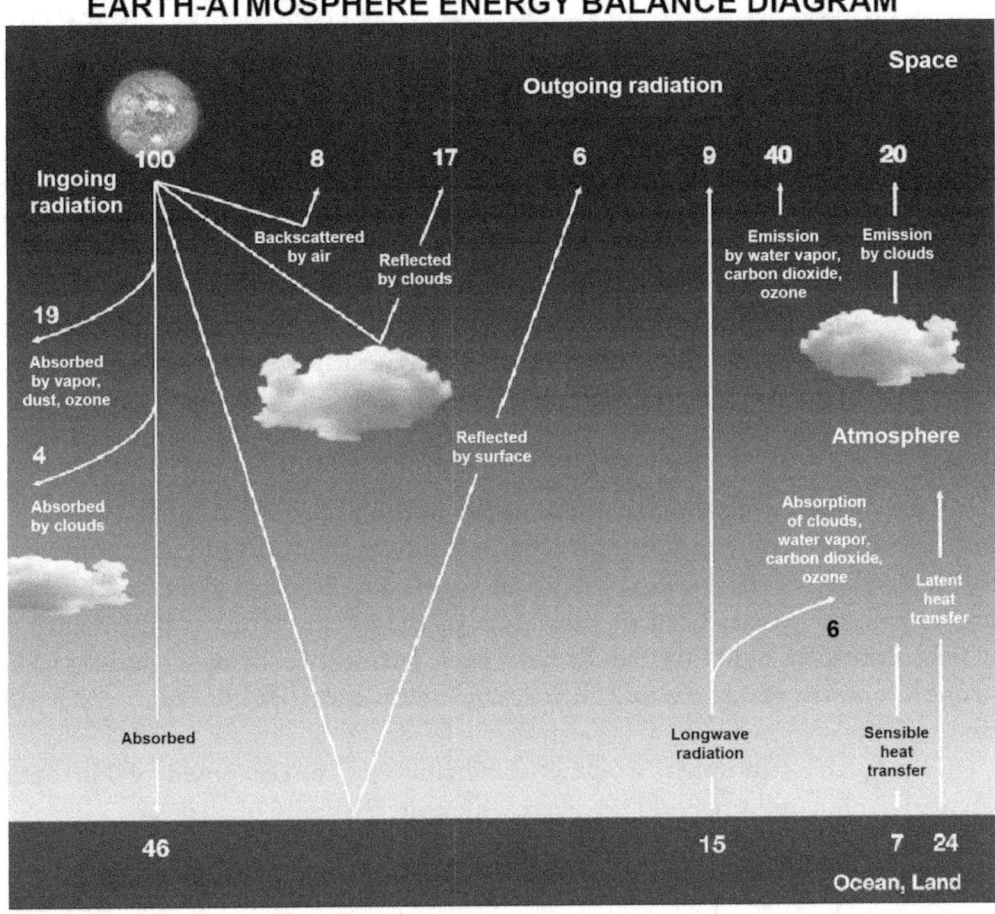

Figure 5.7 – The Earth's energy balance depends on the quantity of solar energy absorbed by oceans, land and atmosphere and the heat quantity re-radiated by the Earth back to space. Source: www.srh. noaa.gov

The natural greenhouse effect[70] preserved, instead, life-suitable long-term conditions on Earth. Despite the innumerable temperature oscillations, the complex dynamics of the atmosphere and variations in the energy balance, climate demonstrated to be tuned in such a way as to be compatible with life even during glacial and interglacial stages.

---

70 The solar energy balance is the amount of solar radiation absorbed by the biosphere, virtually equal to the re-emitted infrared radiation.

This relationship is one of the most astonishing "special conditions" which accompanied the evolution of Earth, biosphere and life. In the absence of GHGs anthropogenic emissions, which made the atmosphere warmer, a new ice age could have started already a few thousand years ago!

The parity balance condition is altered when ingoing absorbed radiation is not equal to the outgoing radiation, which is the reason for the current increase of surface temperature and global warming. The self-balancing capacity of the biosphere, in other words, allows the system to adjust to a surprising variety of life-compatible conditions, despite the occurrence of some extreme events. Our ancestors survived during the coldest condition prevailing in Europe between 50 ky and 20 ky ago and the following deglaciation, which ended around 10 ky ago. The last ten millennia were characterized by a nearly optimal climate trend which helped the farming revolution to start and continue until present.

Climate has been rather favorable to life in the past, but it may not behave so in the near future, due to the unprecedented dimension of the human population, the complexity of the social structure and associated impacts. Another threatening human impact has been the depletion of the Ozone layer, in the absence of which ultraviolet radiation would have been harmful to living organisms. The ozone layer was endangered in Antarctica by the emission of CFC[71] during the past century!

The Earth's atmosphere is a self-stabilizing system, composed of Nitrogen, Oxygen, Argon and Carbon Dioxide reaching the 99.998% of all gasses, Nitrogen alone amounting to 78%. Oxygen, the second biggest component of the atmosphere, is essential to all living organisms and fundamental in the photosynthesis, the process through which CO_2 is converted into organic matter, thanks to the energy provided by the sun. Figure 5.8 shows the current biosphere's energy budget, providing an overview of the energy absorption and utilization, the heat contribution from inside the Earth and human activities.

The Earth's interior provides 31.1 TW, 2/3 of which by conduction (transfer of heat through the Earth's material), and 1/3 by convection (heat from emissions of lava, ashes and hot gases), mostly through the Mid-Atlantic and the East-Pacific oceanic fractures.

---

[71] A new ozone hole, discovered by NASA, opened over the Arctic Ocean this spring as a consequence of a very cold winter. The information was provided by internet (04/10/2011).

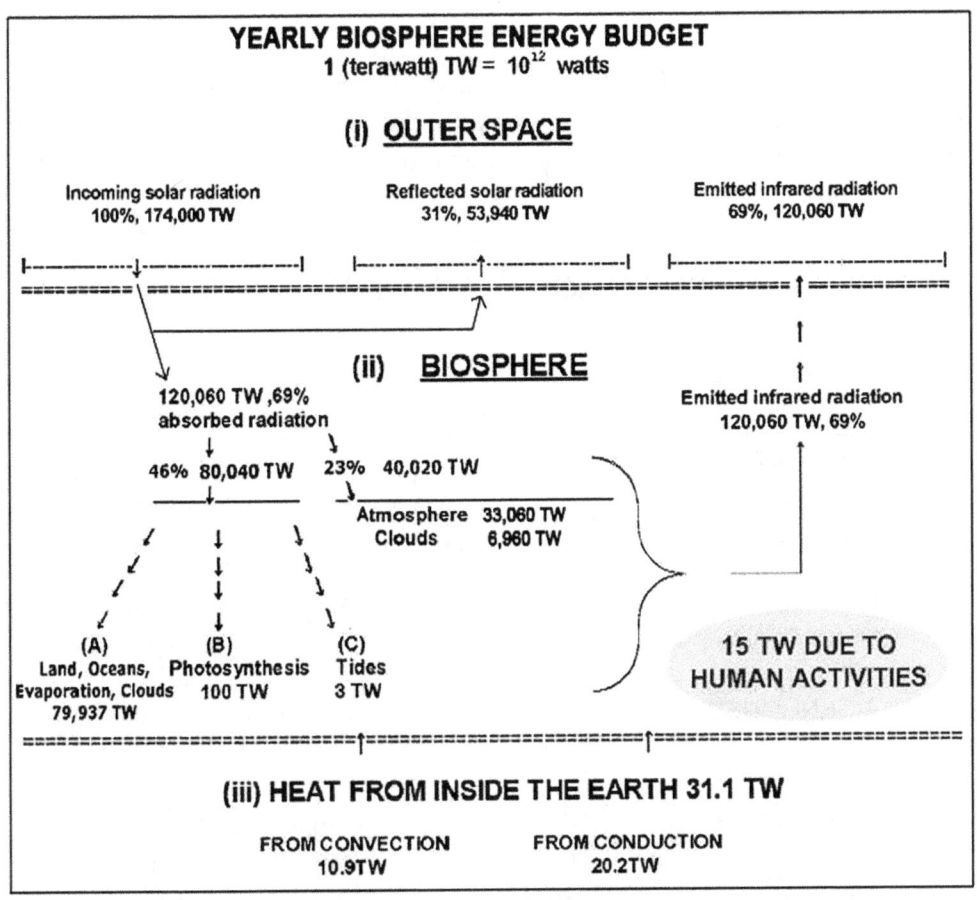

Figure 5.8 – The yearly biosphere energy budget is the result of solar radiation absorption, the heat form the Earth and the unprecedented contribution of 15 TW from human activities. Compared to the absorbed solar radiation of 120,060 TW, the very low human-related value of 15 TW in 2004, by rising at the rate of 1 TW per year, in a two decade time could reach in 2025 the value of 35 TW. This would not significantly alter the energy balance, but certainly increase the greenhouse effect due to the growing concentration of CO_2 which is a most powerful driver of global warming.

THE GLOBAL ENVIRONMENTAL CRISIS

Anthropogenic energy contribution and the global radiative forcing

The human energy contribution to the Biosphere amounted to 15 TW in 2004. Table 5.1 shows the sources and power of the human contribution to the Earth's energy budget in 2004, and the comparison of pre-industrial and 2005 GHGs concentrations.

TYPE OF RESOURCE	Power in TW	%
Petroleum	5.6	37.33
Gas	3.6	24.00
Coal	3.8	25.33
Hydroelectric Energy	0.9	6.0
Nuclear Energy	0.9	6.0
Wind, geothermal and wood combustion	0.2	1.33
Total	15 TW	100

GREENHOUSE GASES (GHGs)	Pre-industrial level (before 1750)	Current level 2005	Increase since 1750
Carbon dioxide	280 ppb	387 ppm	107 ppm
Methane	700 ppb	1,745 ppb	1,045 ppb
Nitrous oxide	270 ppb	314 ppb	44 ppb
CFC-12	0	533 ppt	533 ppt

Table 5.1 – (top) The energy contribution from fossil fuels and other sources in 2004. In the near future, increasing anthropogenic contributions are likely to produce a higher greenhouse effect and faster global warming. (bottom) Emission of major GHGs since 1750. Sources: upper box: USA Energy Information Administration (EIA), International Energy Statistics; lower box : IPCC, Climate Change Report, AR4, 2007.

The fast increase of CO_2 (from 280 to 387 ppm) in 255 years (1750–2005) represents a significant threat to the stability of climate, unprecedented during the past ice ages.

The unusual increase of GHGs — for the first time caused by human activities in the last two centuries — is considered responsible for the current climate change.

Based on the IPCC (2007 Report), out of the 8.8 billion tons of CO_2 emitted into the atmosphere in 2005, 6.5 billion tons are the result of fossil fuel consumption, 1.5 billion of deforestation and 0.8 billion are from other causes. Deforestation represents a major threat to the stability of the ecosystem as a result of slash-and-burn activities and the associated production of CO_2. In addiction a lower absorption of CO_2 is due to the decreasing vegetation cover (caused by the deforestation) and the lower photosynthetic potential. Figure 5.9 shows the components of the radiative forcing and the total net contribution.

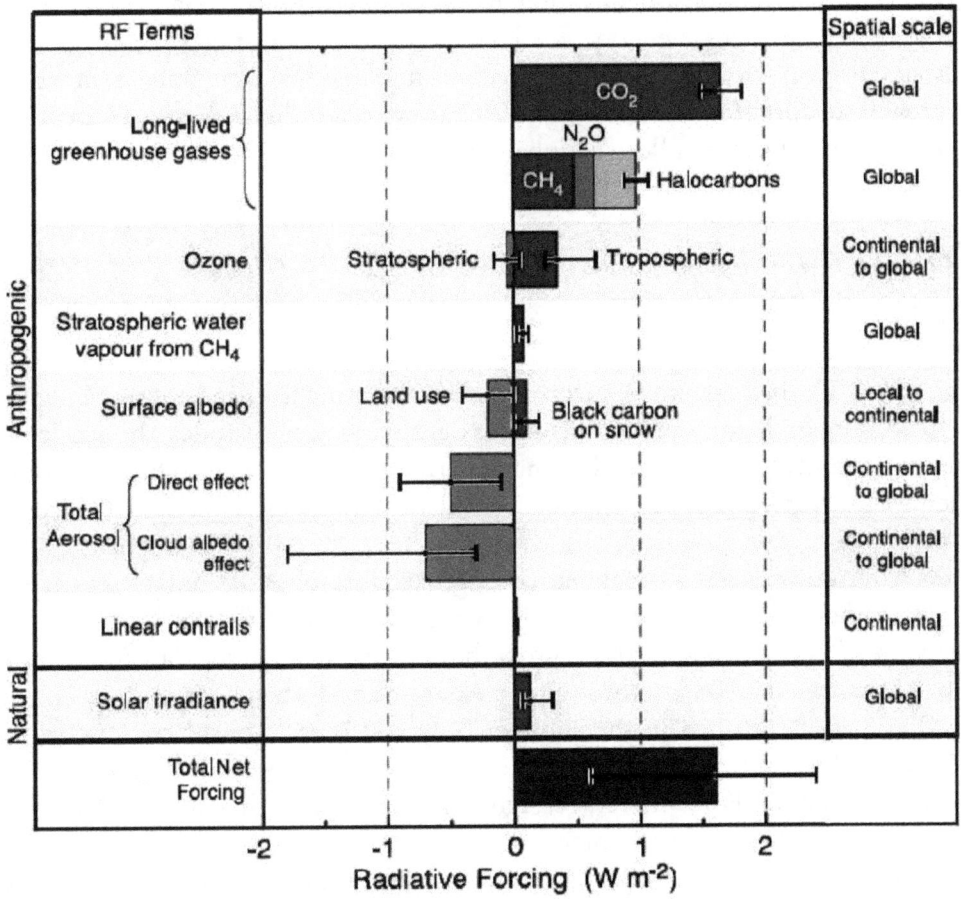

Figure 5.9 – Estimated radiative forcing in 2005. Adapted from: IPCC, Report on Climate Change, 2007.

The major natural greenhouse gas on Earth is water vapour. Human gaseous emissions include CO_2, CH_4, NO_2, halocarbons and tropospheric ozone, balanced by the cooling effects from aerosols and the clouds' albedo. During the last 200 years, GHGs emissions have grown to unusual levels, producing an *enhanced greenhouse effect*, which is considered responsible for the current global warming.

The total net anthropogenic forcing — responsible for increasing global temperature by 1ºC — is estimated at 1.7 W/m², mainly represented by CO_2. A final consideration is that part of the GHGs remains trapped into the atmosphere for centuries, therefore the emission reduction of greenhouse gasses will not necessarily cause a rapid decrease in the mean temperature, but hopefully a slowdown of the global warming trend. Estimates based on atmospheric GHGs concentration — in the absence of changes in energy policies — lead to conclude that the rate of human related emission will rise during next decades making the mean global temperature to grow faster, progressively nearing to 2°C above pre-industrial time. Humanity in sum is already in the middle of a global emergency!

Temperature variation, ice melting, the thawing of tundra and permafrost

The recent 1°C increase in the mean global temperature has been accompanied by a 3° to 7°C rise in the summer temperatures of the northern latitudes, the accelerated melting of land and sea ice, the thawing of permafrost, a greater instability of climate and more extreme seasonal events such as hurricanes and heat waves in the tropical zone.

The Arctic sea ice — mapped through satellite observations — has melted so much that in the summer of 2009 two German cargo vessels followed a new commercial North-Eastern route across the Arctic Ocean to the Pacific, 7,200 km shorter than the 19,000 km through the Atlantic and Indian Oceans.

The passage is open in the summer only and is still considered unsafe. The minimum (September) extension of the Arctic sea ice in 2007, 2008 and 2009 was about 4.78 million km² on average, significantly less than the 7.04 million km² in the 1979–2000 September average[72].

The ice retreat is faster than estimated by any of the 18 computer models used in the elaboration of the 2007 IPCC assessments, and the decline of summer ice is about 30 years ahead of climate model projections[73]. The Arctic region is particu-

---

72 Source: Arctic sea ice extent remains low; 2009 sees third-lowest mark
73 Source: Models Underestimate Loss of Arctic Sea Ice

larly vulnerable to rising temperatures, since sea ice covers are retreating and darker areas of open waters are expanding, thus absorbing more solar radiation. The thick ice covers on Greenland and Antarctica are affected by the rising temperature as well. Unlike the melting of sea ice (which does not increase the sea level), the net melting of land ice and the thermal expansion of ocean waters contribute to increasing the sea level. The monthly changes of the Greenland ice sheet in 2006, suggest a net melting rate of 239 km^3/year[74] contributing in the order of 0.6 mm/year to global sea level rise. An evident effect of the current rise in the mean temperature is the thawing of permafrost in the vast tundra[75] regions of Greenland, Canada, Scandinavia, Siberia and Alaska, the phenomenon being currently monitored and documented. Permafrost holds enormous quantities of carbon in the form of organic matter. According to Ted Schuur, Associate Professor at the University of Florida, when it starts thawing and melt water drains away, the organic matter decomposed by microbes rapidly releases huge amounts of methane and CO_2 into the atmosphere, contributing to further temperature rise[76].

The Thermohaline Circulation Shutdown

If the overshooting level of 2°C is reached in the next decades, two scenarios are possible. The mean temperature could keep rising to 3°C (or more) above the pre-industrial value, the mean global temperature nearing the level of the Eemian interglacial 129,000 years ago (Figure 5.6). Scientists consider this warming scenario possible and catastrophic since it would worsen deleterious phenomena already at work now, like ice melting, hurricanes, flooding, wildfires. Diseases would be spread faster than ever, food production would decline and dramatic migrations of people and species occur. The second scenario concerns the temperature inversion that could take place — when temperature approaches to about 2°C above the pre-industrial value (graph b in Figure 5.5) — due to the shutdown of the global

---

74 Source: Satellite Gravity Measurements Confirm Accelerated Melting of Greenland Ice Sheet
75 *Tundra* is a treeless plain characteristic of Arctic and Subarctic regions. Its soil has a dark brown colour and is rich in organic horizons. *Permafrost* consists of saturated soil, subsoil and even bedrock in Arctic and Subarctic Regions, from a few m to over 1000 m thickness, that has remained frozen for tens of thousands of years. Permafrost contains huge quantities of organic matter, mainly in the form of peat layers, thus when it melts enormous quantities of greenhouse gases (methane and carbon dioxide) can be released.
76 Source: Bad Sign For Global Warming: Thawing Permafrost Holds Vast Carbon Pool

Thermohaline Circulation, Figure 5.10, locally known in the Atlantic as the Gulf Stream.

The Stream brings in fact to the NE Atlantic the heat that makes the winters in the British Islands and Scandinavia much warmer than in the Atlantic USA and Canada. The freshwater entering the North Atlantic, from the melting of the Arctic ice, the Greenland ice sheet and the thawing of permafrost in Canada, are today growing beyond expectation. The discharge of freshwater may become so large to slowdown and stop the sink of dense and salty water brought by the Gulf Stream in the Northern Atlantic, halting the thermohaline circulation.

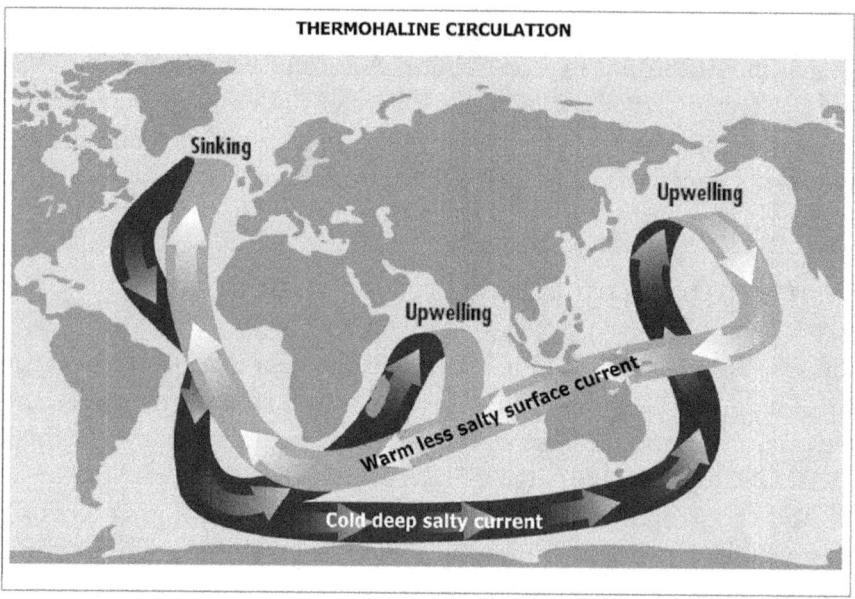

Figure 5.10 – Warm surface water circulation from the North Pacific to the North Atlantic and the deep cold and salty current in the opposite direction.
Source: The Greenhouse Effect And Climate Change

The slowdown of the thermohaline circulation is considered responsible for the sudden cooling of the North-Eastern Atlantic region 14–12 ky ago (Figure 5.6, bottom), known as the Younger Dryas cold stage, which abruptly brought back glacial conditions in the British Islands and Scandinavia. Should a similar sudden cooling occur now — when humanity is nearing the tipping temperatures of the previous four interglacials — a wider accumulation of snow and ice could take place and this would in turn initiate the *albedo* effect, driver of a faster reglaciation stage. New policies aimed at decreasing emissions of GHGs were not agreed in the

2009 and 2010 World Conferences on Climate Change. Extreme geological events associated with climate change are described by Bill McGuire in the New Scientist[77] magazine.

Conclusive Remarks

First, the industrial world's economy is based on the combustion of fossil fuels, a process which generates heat and returns to the atmosphere huge quantities of carbon absorbed by plants and organisms hundreds of millions of years ago. The huge emission of GHGs increases the capacity of the atmosphere to trap heat, therefore driving global warming and the climate change.

Second, the human related change of climate is unusual, if compared to the sequence of ice ages during the last one million years and the associated variation of CO_2 in the range 180–280 ppm. Carbon dioxide content in 2011 has reached 392 ppm, that is a much higher level than in the past 20 million years. Recent studies show that, at the current rate, CO_2 can rise 550–600 ppm by the end of the century, making the average global temperature to rise uncontrollably and accelerating ice melt, sea level rise and effects on atmospheric cycles. Humanity is currently heading towards a climate condition similar — but faster in CO_2 emissions and global warming — to the PETM spike (pages 53–57), which took place 56 million years ago. Humanity today has reached 7 billion people, therefore being trapped in the exponential growth of population, activities and GHGs emissions. Even worst is the time trap: climate change and effects proceed at a much faster rate than humanity's capacity to control the demographic explosion, transform industrial activities and establish a carbon-free society based on renewables.

Third, the growth of unresolved national and international problems, the unpreparedness of governments, the inconclusiveness of international talks on climate change, overstress the human condition and show unequivocal signs of a collapsing civilization.

Fourth, today climate is changing faster than life in general and humanity in particular can adapt to.

Understanding our environmental and human problems is the only way to get the global crisis under control. We have still chances of survival, provided that

---

[77] Climate change affects geological phenomena like sea-bed landslides, earthquakes through loading and unloading of the crust (due to thickening ice covers and unloading associated with melting), volcanic activity (Iceland) and the reactivation of faults. McGuire concludes that a warmer future will also be more geologically turbulent. New Scientist, 2553 Issue, May 27- June 2, 2006

reason and wisdom prevail. The fast growth of GHGs emissions and global warming calls for a common unprecedented change in economy, production and the use of resources, but above all it requires a new approach to organize a society in harmony with environment and the ecosystem, rather than against these two irreplaceable pillars of life.

CHAPTER 6

[A Historical View of Capitalism and Democracy]

The Evolution of Capitalism

The debate over the conflict between capitalism and communism ended with the collapse of the Berlin Wall in 1989, followed soon after by the disintegration of the Soviet Union[78] in 1991 and the failure of its centralized economy. Capitalism today is the global expression of a free-market economy, differentiated in a variety of forms and at the same time a heavily criticized system, driver of profit and instability.

Based on the private ownership of production means (which are operated for profit, in turn reinvested for commercial purposes and further gains), the capitalist economy is deep rooted in its entrepreneurial force considered a fundamental feature of the system.

---

78 The Soviet Union model of communism lasted from 1922 to 1991.

Criticism of capitalism, concerns its basic principles and effects as the unequal distribution of gains, the trend towards monopolizing business and real estate, the interference with democratic systems and government policies. In addition some endemic aspects of capitalism — like the need for low wages and more recently the global attack on environment and the ecosystem — depict a power obsessively affected by the need for an ever growing consumption of goods which openly ignores the limits of the Earth and Humanity.

The fast economic growth of emerging countries and some developing nations shows, in the end, the preference that capitalism has for non-democratic forms of government. From a historical viewpoint capitalism as a system acquired growing economic power and complexity during the past century, becoming inevitably uncontrollable in our overpopulated planet.

Much before the farming revolution[79] humans had already developed the concept of property in terms of possession of land, hunting tools, captured preys, and the unquestionable right to preserve it as a precondition for survival. Property, authoritarianism and slavery during the last eight millennia were the bearing structures of great empires. The first legal definition of property belongs to the Corpus Juris Civilis[80], introduced by the Emperor Justinian from 529–534 A.D., which defined it as *"the right to use and abuse of ones own things, within the prescriptions of the law"*.

The absolute possession included "abuse" as a right and, as J.K. Galbraith noted in his book *"Economy in Perspective"* (1987), excluded any possibility of interference from outsiders. At the same time "prescriptions of the law" since then have implied limits beyond which that right was considered illegal. During the past centuries, however, the awareness of possession prevailed. Recent progress in the recognition of human rights and democracy — as the unique system in which industrial production and market economy can coexist — is promising, but not advancing as fast as it would be necessary.

Merchant capitalism as the oldest type of historically documented commercial system dominated the global market until the 14th century in Europe, mainly practiced by Mediterranean countries and the Northern European nations. The discovery of the Americas in 1492 fostered that form of capitalism opening the way to the creation of great commercial empires which then expanded in the direction of the New World and the Far East. The biggest change in the structure of the merchant capitalism occurred, however, in coincidence with the industrial revolution (1800s)

---

79   The farming revolution started 10,000 years ago
80   The Body of Civil Law

and the new technologies for the construction of big ships and railways which revolutionized the commerce worldwide.

During the last few decades, a new phase developed, that is "supercapitalism". Due to its growing tendency to drive politics and the economy, decide national and global choices, control the public sector and reduce in some cases democracy to a merely bureaucratic function, supercapitalism grew beyond limits. Since the beginning of 2000s its unprecedented expansion heavily endangered the global economic system.

Democracy as an Unfinished Process

Democracy was conceived 2500 years ago in Athens at the time of Pericles[81]. The experiment, never attempted before, saw the participation of Athenians (and their resistance to the invasion from the Persian Empire). The Greek democracy lasted 150 years ending in the 4th Century B.C.

A resurgence of democracy occurred as a consequence of the Age of Enlightenment (1700–1800) in Europe when the concept of a modern democracy unfolded on the basis of the French Revolution (1789–1799) and the motto "Freedom! Equality! Brotherhood!".

Marx and Engels through the *Communist Manifesto* (1884) developed their criticism to capitalism opening the way to the 1917 Russian Revolution, followed in 1922 by the creation of the USSR. Later on, after World War II, the split of Europe and the construction of the Berlin Wall in 1961, marked coexistence of capitalism and democracy in western economies only.

Democracy-based nations during the period 1950–1980 made a considerable progress in economic and social terms, showing that common shared rules were essential to governance and instrumental to a proper functioning of the market competition.

With the collapse of the Soviet Union in 1991 a long and dramatic historical period ended. Europe was re-united and the general hope was that finally capitalism and democracy could establish a new trend of peace and cooperation inside and

---

81 The Athenian democracy started around 500 BC. The new system was established under the reforms of Cleisthenes who turned the Government from an oligarchy into a democracy which developed during the fifth and the fourth centuries BC. The basic principles were equal rights for the citizens and a control system through which no person or no group could become exceedingly powerful. Pericles was the most famous leader of the Athenian democracy. Under Alexander the Great, who died in 322 BC, the Aegean territory was included in his empire, thus ending the first experiment in "Democracy".

among countries. The European Union then expanded to 27 independent States with half a billion inhabitants, that is the 7.3% of the global population, and the 20% production potential of the global GDP PPP[82].

The Union is today facing uncertainties and economic problems, mainly due to historical and organizational differences among member Countries. During 1990s BRICS economies (Brazil, Russia, India, China, South Africa) began developing rapidly, showing that their combined efforts could eclipse the traditional economic trends of rich countries.

Since 2007 through 2011 the economic crisis in USA and the EU turned into a major global problem. The increasing foreign debt of the USA and some EU nations, the unprecedented power of international investors, and the rapid economic growth of emerging economies, by the end of 2011 slowed down the growth of rich western countries.

The central point of capitalism is that money goes where investments are safer and more profitable, thus indirectly isolating economies whose inadequate control institutions have allowed the income and spending of their nations to diverge excessively.

In this view Supercapitalism has become a supranational power no longer limited by the traditional boundaries of countries, more powerful than governments and capable of determining national and international crises.

On the other hand, the foreign debt of some EU countries has demonstrated the unpreparedness of the national authorities in setting up an efficient control system and the difficult management of a Union composed of non homogeneous countries.

The democratic system adopted by EU Countries, however, is still a reference experiment that, if halted, would destabilize the global economy and probably trigger a recession process worldwide.

Historical Views by Smith, Malthus, Mill and Clausius

The theory of capitalism is based on **Adam Smith's** book *The Wealth of Nations*, published in 1776, which is coincidentally the date of the American Declaration of Independence. Smith's ideas on the free market economy were soon adopted in Europe and the USA, then by other countries and very recently, after the collapse of the Soviet Union, by Eastern European Nations, Russia and unofficially by China.

---

82 World Economic Outlook Database, September 2011 Edition , IMF

A HISTORICAL VIEW OF CAPITALISM AND DEMOCRACY

According to Smith, the economic system is stimulated by people's interests and egotism, price competition and the distribution of income resulting from production and commerce. All together these properties of capitalism are supposed to promote prosperity. The personal interest, which provides the energy to the process, is seen as an invisible hand which leads individuals to seek what is best for themselves.

During Smith's era (1723–1790), the belief that the economy could grow continuously and find new resources in the environment was understandably widespread. The world was sparsely populated during the middle of the 1700s (about 1 billion people) and vast forests extended across the planet.

Natural resources and the unexplored lands of the Americas seemed unlimited. Immigration to USA during the last two centuries significantly fostered population increase, generating economic growth, job opportunities, the creation of a powerful industrial sector and cultural-ethnical controversies as well.

Figure 6.1, depicts capitalism as a tornado, which cannot obviously expand forever.

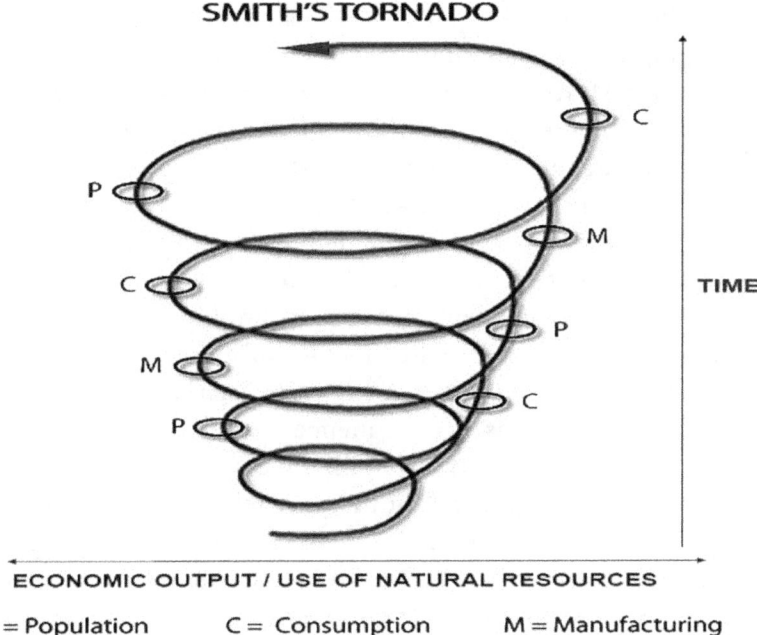

P = Population   C = Consumption   M = Manufacturing

Figure 6.1 – Smith's Tornado depicts a growing economy supported by population rise, a consumption based on natural resources and production, followed by a further manufacturing and population increase. Spirals in the last decades expanded beyond the size of the ecosystem. Adapted from Towards an Environmental Political Economy, Essays in Philosophy, A Biannual Journal 6 (2), June 2005.

The sequence shows the population growth (P) which increases the consumption of food and goods (C) through the market demand, which in turn stimulates further manufacturing (M). Between 1850 and 1950 millions of people migrated towards the Americas in search of a better life conditions and freedom. Smith's theory, has been recently fostered through the adoption of a free market-like economy by China and other former Soviet Union countries. Despite the undeniable success, an unlimited expansion of the economy is absurd, since it would imply the possibility of creating matter and energy (that is resources).

The concept openly contradicts the principle of conservation of energy in physics which states that matter and energy can neither be created nor destroyed. Demographic growth and profit are the driving forces of the system, these two components determining the tornado shaped process which develops into an automatic spreading trend in the form of spirals of increasing dimensions through time.

The gross world product over the last 2000 years can be represented as an irregular conical helix, in which spirals grew very little versus time until 1800, when a rapid expansion of the economy abruptly made spirals to grow 234 times bigger during the last two centuries[83].

The demographic problem of development was anticipated by the English economist **Thomas Malthus** (1766–1834), who pointed out the incompatibility between the exponential growth of the population and the linear increase of food production as a result of a linear growth of arable land. Malthus warned that the quantity of cultivable land was limited and that in the future, part of the population would have faced insufficient availability of food. Food production instead has greatly increased since the industrial revolution, as a result of mechanization in agriculture, new advanced technologies, the expansion of irrigation and the use of fertilizers and biocides. Today, however, over 1 billion people are undernourished and due to the food market conditions in 2010 it is reasonable to expect that, as a consequence of the further growth of population to about 8 billion in 2025, their number will increase dramatically in poor countries. The Malthusian view, although not felt as a major problem, is still a nightmare for two reasons (i) the unlimited growth of food production is absurd since unsustainable in the mid-term due to the deterioration of the environment, the decline of soil quality and the pollution of freshwater and, (ii) the yearly food quantity in the global market is not equally available to all, growing quantities of food being imported by rich and emerging countries and used also for fish and animal farming. By the end of this decade, the economic growth of

---

83   The estimate of GDP was done by J. Bradford DeLong, Department of Economics, U.C. Berkeley: Estimating World GDP, One Million B.C. – Present

emerging Countries whose food consumption is increasing faster than production, the impoverished quality of arable land, the rising demand for freshwater, the increasing cost of non-renewable resources, will cause food availability to reach the upper limit.

John Stuart Mill (1806–1873), a British economist and civil servant, stressed the idea of individual freedom in contrast with the control by State institutions. Mill is considered to be the first economist with a global view of the limits of the planet and the advocate of the "stationary state of the economy" as the only solution for a growing society.

Some of his most advanced views concerned (i) the inconsistency of unlimited growth and, (ii) the concept that a stationary condition of capital and population does not imply a stationary state of human material and spiritual progress. Mill anticipated more than a century ago disadvantages and risks of an ever growing uncontrollable economy. Finally he was a strong promoter of economic democracy within the free market, with the labour force entitled to elect management members, in contrast with the authoritarian behaviour of private companies. Mill's prophetic views on capitalism, the effects on the environment and advantages of the stationary state brought him to conclude "*...I sincerely hope, for the sake of posterity, that they will be content to be stationary, long before necessity compels them to it*"[84].

The German physicist **Rudolf Clausius** (1822–1888), argued that the huge coal deposits which had been discovered[85] were being irresponsibly consumed, suggesting a larger use of renewable resources such as forests[86]. The problem mainly concerns today oil and gas and coal, which is still available in large quantities. Coal is the biggest producer of CO_2 and a driver of global warming, growingly used in the production of electricity.

---

84 J.S. Mill *"Principles of Political Economy"*, (1848) Book IV, Chapter VI.

85 Coal deposits are due to the accumulation of large quantities of vegetation. Plant remains over geological time are compacted, chemically altered and covered by sediments which prevent oxidation.

86 Over a century ago, Clausius understood that renewable natural resources (such as forests) could have been used without exceeding the natural rate of regeneration, since non-renewable resources, (only coal at the time) would have run out sooner or later. This scientifically logical statement was made in the 1800s when the amount of coal in the world was not yet known and the use of oil, gas and nuclear energy would have developed much later. Clausius did not make any predictions in terms of time to exhaustion, but he clearly spelled out the concept that man was wasting non renewable energy.

Criticism of Lobbies and the Traditional Market Economy

The first opposition to capitalism came from socialist thinkers during the 19th century and the debate widened during the 1900s as a consequence of the Russian revolution in 1917. With the "cold war", after 1950s, the contrast between communism and capitalism rose again, supported by conflicts in Africa, Asia, Central and South America.

Criticism to capitalisms and communism is a central point in some of the books that **Herbert Marcuse** (1898–1979) wrote between 1964 and 1969. The philosopher, highlighted the dramatic effects of the cold war and the trend of capitalism towards the concentrations of capital, the huge unnecessary consumption of resources and the consequences on environment and society. In his famous book *"One-dimensional men"* (1964) Marcuse warned that humanity was abdicating to the reasons of the advanced industrial society, by accepting the unlimited exploitation of natural resources and a comfortable, democratic unfreedom.

Sound criticism on the *affluent society* is the core of several books written by **Kenneth Galbraith** in the 1960s and 70s, based on the concept that a progressively rich society carries better living conditions, and critical aspects such as overconsumption, environmental destruction and the interference of lobbies at the political decision-making level.

In his book *"Economics and the public purpose"* (1973) Galbraith stated that economic development is essential to private enterprises, warning as well that negative consequences on environment, the atmosphere, water and life should not be accepted as externalities (unavoidable side effects of the economy), since distorting the market competition. The remedy according to Galbraith was that governments should invest public money in research and development, limiting industry-related impacts as much as possible. This was the most advanced view during 1970s, when population grew from 3.8 and 4.5 billion people. A major aspect of his criticism was the interference in politics of corporations, companies and lobbies and their attitude to trespass democratic rules.

Herman Daly[87] in his 1996 book *"Beyond Growth"* provides the best example of criticism towards the conventional free-market economy and the idolatrous orthodoxy that hamper the understanding of the biophysical limits to growth and the ethical-social limits that render it undesirable. Sustainable development in the form of a stationary state of the economy is recommended as a solution to the indefinite

---

87 H. Daly is an Ecological Economist, Professor at the "School of Public Policy", University of Maryland, USA

growth which is impossible and risky for a common survival. His views challenge the traditional economics, highlighting the dangerous trend of externalities and the inconsistency of the GNP, which are formidable drivers of resource destruction and environmental impacts worldwide.

J. Rifkin considers the current global impact to be the inevitable trend of capitalism and a consequence of 20th century activities. In his book *"The end of work"* (2002) he highlights novelties and trends:

- the *information revolution*, which has enhanced the society's capacity of substituting man with highly-sophisticated machines

- the *beginning of the access era*. Emerging economic activities are changing the foundations of capitalism. Property will continue to exist, but will be replaced by distribution networks, which are growing rapidly. The access era is a new way of managing life and the economy, by accessing services, rather than owning goods.

- the *transition from markets* to networks concerns the restructuring of the productive world (already underway), based on computer science, automation, robotics, renewable energies and efficiency.

- the *increasing unemployment* as a consequence of the modernisation process. A sustainable global economy needs to change the society by creating more job opportunities as the only alternative to social unrest. Jobs declined in the industrially advanced World between 1960 and 1990, even though industrial production had risen and automation and information technology as well.

Questioning the traditional way of using natural resources and the concept that human history is necessarily based on material growth, his radical proposal is a *"global and stationary economy"*. A sustainable global economy should result in a high standard of living, by adjusting production and consumption levels to nature's capacity to recycle waste.

Rifkin defines this emerging economic process as *"cultural capitalism"*, in which property is maintained but used in a new context, that is in favour of humanity.

Robert B. Reich[88] in the book *"Supercapitalism, The transformation of business, democracy, and everyday life"* (2007) provides an overall view of the economy and the power of lobbies in the US. Salient points of his vision concern the transition to *"supercapitalism"* between 1970 and 1990 and the globalisation that was accomplished by the end of 1990s.

Naomi Klein[89] in her 2007 book *"Shock Economy"* describes the evolution of *"supercapitalism"* and a new wave of business centred on disasters, like the flooding that devastated New Orleans (August 2005).

The role of the Chicago School of Economics and the suggestions by the economist Milton Friedman were instrumental to the transition of China towards a free market economy. Chinese central Government brought in less than 20 years the country to the level of the biggest industrialized global powers. A widespread deregulation, the absence of labour unions and the presence of a centralized political power, enabled China to do better than western economies in terms of GDP growth and production.

The book *"Common Wealth"* (2010) by **Jeffrey Sachs**[90] provides a critical view of the American capitalism in terms of environment, impacts and the need for an international approach to global problems.

Many other writers criticised capitalism, in a number of cases not even mentioning it, but definitely considering it as an almost untouchable system inevitably leading to a concentration of capital which in turn limits the free competition. At the moment capitalism seems to limit democratic rules, now more than ever being source of discontent and social contrasts. The way global situation is evolving now is a larger consumerism, a weaker public sector, and a subtle but growing deregulation.

Emerging Asian Powers and Global Effects

The first premonitory signs of a change in the global economic order manifested during the 1980s, when private companies from the industrialised world began progressively moving their industrial production in Asia (India, China, Vietnam, the Philippines) and other countries. Western industry made huge investments

---

88 R. Reich, Professor of Economics, University of California, Berkeley and Minister under Clinton's Presidency.
89 Naomi Klein is a Canadian journalist, actively involved in political analysis and criticism of capitalism.
90 Jeffrey D. Sachs, Director of the Earth Institute, Columbia University.

abroad where labour wages and production costs were lower, deregulation adequate and environmental protection rules nearly absent.

Lower production costs by the Japanese industrial car-sector in the 1970s had already represented an example of the high quality and competitive manufacturing capacities in the Far East. Table 6.1 shows the trend of GDP (PPP)[91] in trillion US$ for the major economies.

GDP (PPP) in Trillion US$ - Estimates by IMF							
	1990	2000	2010	2011	2012	2014	2016
WORLD	23.530	39.900	76.285	78.290	82.912	93.121	105.545
USA	5.800	9.951	14.657	15.227	15.880	17.223	18.807
EU			15.170	15.608	16.151	17.331	18.722
JAPAN	2.326	3.213	4.309	4.417	4.571	4.849	5.145
CHINA	0.910	3.013	10.085	11.174	12.407	15.289	18.975
INDIA	0.749	1.582	4.060	4.447	4.863	5.861	7.106

Table 6.1 – GDP PPP overview 1990-2016. China is expected to reach US and EU economies by 2016. The debate over 2016 data estimate involves a variety of geopolitical aspects, including the traditional leadership of America and the future of globalization.
World GDP PPP data concerning 1990, 2000, 2010 are from GDP; PPP (US dollar) in World (Trading Economics); GDP PPP 2010 European Union data are from Economy of the European Union (Wikipedia); all other data are from IMF: List of countries by past and future GDP (PPP) (Wikipedia)

USA and EU should carefully consider the expected trend between 2011 and 2016, which implies the global leadership of the emerging economies (BRICS) and the unavoidable decline of the western powers. Europe and other nations should preserve their political and social levels by finding the way for free market and democracy to coexist constructively, being complementary to one another.

Conclusive Remarks on Capitalism and Democracy

The concept of a modern advanced society is deeply rooted on the optimal combination of democracy and the free market economy as it appeared to be in the USA between 1960 and 1970. *The New Industrial State*, by J. K. Galbraith,

---

91   GDP PPP: Gross Domestic Product, People Purchasing Power.

published in 1967, illustrated the concept of an advancing capitalism, based on big corporations, new technologies and organizational skills.

The Columbia University protest in 1968, demonstrated, instead, the weakness of democracy and the connection of Governmental Institutions with the free-market economy. The current debate over these two pedestals of the modern society implies the reach of a new consensus. In particular:

- market economy (based on a "supercapitalism" in the last decades) has progressed fast, thanks to results and gains by the private sector, but also it has highlighted the inconsistency of control institutions and the absence of international rules. The current crisis of Western powers and the production rise by the BRICS can foster a global recession

- democracy at the same time needs to improve at the national and international levels. The current crisis dramatically highlights the inconsistency of the social and structural organization of some democracy-based western nations, the foreign debt of which grows faster than it does in other less advanced countries. Unless national and international agreements, are implemented, lobbies will undermine democracies by limiting their governance potential

- globalization, as it proceeds now, shows that benefits are uneven, prosperity remains unequally distributed and ecosystems decline.

Capitalism knows the direction to follow, the same way as water does due to the force of gravity. By contrast democracy is a human cultural aspiration. Therefore — like flowers in a garden need patience, knowledge of their habitat, water and fertilizers — democracy should be grounded on public education, the separation of powers, the mutual respect of Institutions, the cultural level of politicians and policy-makers. For the first time humanity has to deals with the optimization of a global free market economy and democracy, as the governing system of the spaceship Earth, with the twofold responsibility of preserving the ecosystem and adapting the free market to an optimal and sustainable survival of 7 billion in 2011 and 8 billion within 2025.

CHAPTER 7

Economic Growth and Sustainability

The 1987 Brundtland Report and United Nations Involvement

Introduction

In the 1960s, approximately 15 years after World War II, the decolonisation process was underway and new countries — often rich in natural resources — were emerging with great hopes of economic development. However, as early as the 1970s, this optimistic vision of the future began to fade out and the prevailing trend became that of a divided world of rich and poor countries. A number of developing nations were able to improve their condition, but the large majority, in

Africa and other continents, were at a standstill and in a social and environmental declining trend.

In the 1970s, a sudden diffusion of environmental literature took place and a number of environment-related problems emerged during the 1972 First World Conference on Development held in Stockholm.

The debate concerned overpopulation, development, resources and the environment. Differences and suspicions emerged and the growing divide — between rich countries endowed with resources and technological capabilities, and poor countries in extreme difficulties — became an international issue. A major problem was that rich countries largely controlled oil prices and other natural resources, thus enjoying growth in industrial production and GDPs, often at the expense of developing nations.

In 1983 the Brundtland Commission, formally the World Commission on Environment and Development (WCED), was created and in 1987 the report *"Our Common Future"* launched the proposal of a "sustainable development" that would not deprive current and future generations of the possibility to fulfil their basic needs. In the following years the debate involved a growing number of nations and a sequel of international Conferences was held[92] to stimulate countries to understand the growing threat of the global environmental crisis and associated problems.

The full involvement of the UN and the participation of governmental and non-governmental organizations (NGOs) helped to highlight the need for a sustainable development process. The 1992 Rio Conference reaffirmed the commitment by the United Nations to the environment and put forth a proposal focusing on a sustainable development that recognised the holistic nature of the Earth. Among the Principles of Rio Declaration on *Environment and Development*, two were fundamental:

- Principle 5: All States and people shall cooperate in the essential task of eradicating poverty as an indispensable requirement for sustainable development, in order to decrease disparities in the standards of living and better meet the needs of the majority of the people in the world;

- Principle 8: To achieve sustainable development and a higher quality of life for all people, states should reduce and eliminate unsustainable patterns of production and consumption and promote appropriate demographic policies.

---

92   Earth Summit in Rio de Janeiro (1992), the Framework Convention on Climate Change in Kyoto (1997), the World Summit on Sustainable Development in Johannesburg (2002) and the Conferences on Climate Change in Copenhagen (2009) and in Cancún (2010).

In 2002 the World Summit on Sustainable Development, held in Johannesburg, concluded with a commitment to important issues:

- eradicating poverty and halving the number of people lacking access to safe drinking water and basic hygiene by 2015;

- supporting a decennial programme towards a sustainable consumption of resources and the related industrial production;

- reducing greenhouse gas emissions to control global warming.

The 2005 *Millennium Ecosystem Assessment* (MA) Report by the United Nations depicted a global condition of ecosystems and sustainability with very few signs of improvement. Despite remaining the only alternative in the first decade of the III millennium, Sustainable Development is still far from being an operative course of action.

In 2010, 18 years after the Rio Summit and 23 years after the Brundtland Report, an ethical vision of international problems has emerged, but new policies and agreements remain ignored. Since 1987 the condition of the environment has largely deteriorated and if business continues following its "as usual" trend, beyond any doubt the ecosystem will be even more dramatically affected in the future.

How long can humanity ignore environmental degradation and fail to recognize the economical and social drivers behind it? This issue, whether we head towards sustainability or self-destruction, is entirely tied to our species.

The Brundtland Report

The 1987 Report of the Brundtland Commission was based on the conditions of the Earth and humanity in the mid-1980s, but the basic picture already encompassed many of the present problems:

- the degradation of terrestrial ecosystem, population and economic growth, consumption of resources, all together interact to cause environmental deterioration. The relationship between ecology and economy is evident and the destructive potential of the latter upon the environment is indisputable. There are national and international pressures in favour of an excessive consumption of resources. In the case of poor nations, local natural resources are not entirely used to promote the development of population, a significant component being kept for the restitution of accumulated debt. Due to this, some developing countries — exporting large quantities of non-renewable resources — are facing growing poverty. Natural resources should constitute the basis for human survival and better living conditions for all. Hence the Report recommended an efficient management, based on a sustainable development, so as to satisfy the needs of both present and future generations who are supposed to live in conditions of equality and ecological health

- the excessive use of natural resources, although producing economic growth, profits and an increasing GDP, threatens future generations, which will inherit fewer resources and greater environmental debt. The report states that humanity is borrowing environmental capital from future generations, without any intention or possibility of returning it

- a new era of economic growth-based on policies that preserve the biosphere's resources and ensure the survival of humanity in its entirety – is considered indispensable for a sustainable progress

- growth can foster the reduction in infant mortality, a growing life expectancy and increase the rate of literacy

- global military expenditure has reached 1 billion US$ in 1987, subtracting resources from human well being and increasing local conflicts.

Negative aspects of the economic growth included: (i) the increasing gap between the rich and the poor, (ii) million hectares of arable land worldwide are degraded by desertification each year, (iii) 11 million hectares of forests are annually converted into agricultural land, barely fertile enough to sustain production for the local population and, (iv) CO_2 emissions from fossil fuels and the associated global warming.

Sustainable development, moreover, implied that in order to satisfy the basic needs of the global population at the time, rich countries should have adopted lifestyles compatible with the ecological limits of the planet and the economy should not have ignored the importance of ecosystems. The report correctly stated that development is sustainable only if the dimensions of the population compares to the productive capacity of the ecosystem. However, no indication was provided on which way to follow economically and socially. Although terms "development" and "growth" were used interchangeably, the ambiguity of concepts is understandably related to the time in which the report was conceived. In the end, *Our Common Future* can be credited with highlighting sustainability, environmental and social problems, stimulating nations and international institutions in the direction of a comprehensive vision of environmental issues, and keeping the debate alive at an international level. Despite manipulations by politicians and economists in considering "sustainability" synonymous with "economic growth", *sustainable development* remains the only possible scenario for present and future generations. The Report highlighted the on going change from a compartmentalized world in which activities and effects were held within nations, to an open world in which compartments had started dissolving and interlocking crises (environmental, economical and social) had began expanding worldwide. In 1987 according to United Nations data and projections, the planet hosted 5 billion people, that is a population expected at the time to grow during the next century and stabilize between 8 and 14 billion people. The majority of the increase was supposed to take place in the poorest Nations and concentrate in already overcrowded cities. The view of a massive population, beyond 9 billion in 2050, is considered today unrealistic.

The Sustainable Development by Herman Daly

A voice in the darkness

Criticism of the current economic orthodoxy and a high level contribution to the understanding of the ecologically sustainable development is provided by Herman Daly's[93] several books, among which *Beyond Growth* (1996). The Author criticises the actual fideistic approach to the free market by modern economics, describes specific distortions, the currently inadequate national and international legislation and outlines the pathway for a new era of development.

---

93   Herman Daly, Professor of Economy at Louisiana State University, co-founder of Ecological Economics, worked as Senior Environmental Economist at the World Bank, Professor at the Maryland School of Public Affairs since 1994, author of books on ecological economics, the environment, sustainable development and population. Daly, born in 1938, has been writing and teaching for half a century, that is the most critical period during which the environment crisis, ecological disaster, affluent society, extinction of ecosystems, depletion of resources, and globalization have entered the everyday life.

According to the Author, our World is affected by an environmental crisis, driven by the population growth, the depletion of non renewable resources, impacts associated with human activities, the main reason for this being an economy based on growthmania. A progressively distorted legislation and a social organization no longer adequate to the survival of humanity undermine the global equilibrium. Unlike other books in economics, *Beyond Growth* excels in clarity, wisdom and logics.

Daly's vision implies a multidisciplinary holistic approach which opens the reader's mind to the complexity of natural and social systems and their interactions. The stationary state — which is a long term evolutionary process — is considered the feasible alternative to the growth economy, which is globally responsible for the environmental deterioration and the pollution of the biosphere. The criticism to the free-market economy and the social organization, both grown uncontrollably beyond the physical limits of the Earth and the ethical and social limits of the system, are central to Daly's advanced view of sustainable development. Conventional ideas on the growth economy and the behaviour of the International Banking System, are criticized for their absurd mechanisms and the associated problems at national and international levels. Unlimited growth is charged with the responsibility of making poverty to increase further in developing countries, being indirectly the cause of civil wars and environmental deterioration.

Threats imposed by our society to the survival of life, the dominating anthropocentrism in which the Western economies grew up, and the contradicting financing policies of the WB and other Institutions are considered by Daly drivers of the current state of the world. Traditional views and economic policies of some International Organizations are based on an unscientific approach, which neither recognizes natural and human limits, nor the conflicting measures taken in support of developing countries. The picture on how the free-market currently behaves, where it is leading to and the environmental disarray that generates worldwide, are considered worrisome and astonishing.

Orthodoxy and ignorance

According to Daly the orthodoxy of the current economic process is deeply rooted in the academic domain, a source of an uncritical indoctrination of students. Economics, unlike the disciplines in the natural sciences domain, is considered a system isolated from the natural environment, which is the only "source" of non-renewable and renewable resources, essential to the production process and at the same time the "sink" of generated waste.

The sustainable development concept on which the Brundtland Report is based, should have been more appropriately named "ecologically sustainable development" and involve population, resources, industrial production and the environment. Major events during the past decades — including wars, genocides, social and political upheavals, environmental and human degradation, advances in science, technology and human rights — helped people to understand the current crisis, but politics has basically ignored the possibility of a sustainability process. The obvious question is: why the present economic process remains unchanged, even in the presence of so many associated impacts on environment and humanity? The answer is that while human progress and advancements in the science domain are fast, politicians, policymakers and traditional economists are slow or reluctant to change. The current economic process is considered self-conservative, based on obsolete academic concepts inherited from the past and supported by the industrial revolution which, in the end, was rooted in a simple pragmatic view: the *business as usual* approach (BAU). Daly highlights that is time now to understand that the economic system is:

- profitable for progressively fewer and richer individuals or companies

- dangerous because of its inevitable environmental impact, the depletion of resources and the unstoppable decline of the ecosystem

- no longer leading to an increase in people's welfare, although the economy is continuously expanding

- not supported by a scientific basis indispensable in an overcrowded and highly polluted world.

Concepts and definitions

According to Daly, two factors in the domain of sustainable development need to be considered:

- the first based on the concept of *utility*, that is the usefulness of the process to present and future generations, which must be the same

- the second based on the concept of *throughput*, the amount of material processed in the unit of time, which should be the same for present and future generations, implying that the capacity of the ecosystem to support humanity should not decline.

Daly is in favour of the second definition — which is central to a society based on renewable resources, since the depletion of the others is inevitable — and the basic reason is that throughput can be measured. The reference disciplines for a correct understanding of sustainable development are:

- *Ecological economics*[94] — according to the definition of the economist Malte Faber — puts emphasis on human involvement in the ecosystem, and sustainability related to nature, justice and the time domain. It focuses on environmental change and intergenerational equity, with sustainable development as the basic approach for analysis

- *Environmental economics* — based on the definition by the National Bureau of Economic Research[95] — undertakes theoretical studies of the economic effects of national or local environmental policies around the world, dealing with costs and benefits of alternative policies in air and water pollution, toxic substances, solid waste and global warming

- *Steady-state economy* is a condition consisting of the renewal (not the growth) of population, and the depreciation and replacement of goods based on qualitative improvement of production technologies

- *Growth economy* defined as the increasing physical and energy throughput supporting the activities of production and consumption of goods and commodities. As such it depicts a never ending dynamic process, based on the unscientific concept of unlimited resources.

---

94 Founders of Ecological Economics are K.E. Boulding, N. Georgescu-Roegen, H. Daly, R. Costanza and others.

95 NBER Is a private, non-profit US research organization.

The importance of the carrying capacity and the Plimsoll Line

The biosphere's carrying capacity is described by Daly through the concept of the Plimsoll line, which is the maximum safe waterline of a ship, marked on its sides. If the ship is loaded and that load is well balanced in quantity and weight distribution, the Plimsoll line marks the loading limit. When the water level reaches the Plimsoll line, the ship is at its maximum bearing capacity and still in safe conditions, but additional loads, even if correctly distributed, will necessarily cause the safety line to be submerged.

Despite being the most sophisticated component of the Earth, the biosphere[96] is vulnerable and, as a consequence of the recent and abrupt human development, is subject to the risk that humanity trespasses the planet's carrying capacity limit. The interaction of physical boundaries, natural cycles and the overall dynamics of the biosphere on the one side and anthropogenic activities, impacts and resources consumption on the other, made humanity to reach the upper limit of the Earth's sustainability condition in 1970s. Moreover people and resources on Earth are not evenly distributed and the rich one third of the world's population consumes more energy and matter than the remaining two thirds. Finally there is no wiser commander of a ship, than the one with the appropriate scientific background and technical skills to keep the system under control on the basis of limits to be observed for a safe navigation. The Earth spaceship has no commander, but political leaders (of nearly 200 Nations), the majority of which with a background in the traditional growth economy! A group of conventional economists, according to Daly, *"for much of the latter half of the twentieth century..... has enjoyed unparalleled influence over the course of the World development. Despite considerable successes, there have been numerous glaring failures"*. In sum, under the current trend of human impacts and the sequel of natural phenomena, permanent safe and stable conditions of the ecosystem are not easily achievable. Humanity, however, has still the chance to control climate change and pollution, by halting GHGs emissions, developing renewable resources and protecting environment and the ecosystem. Science and technologies can help a rapid transition to a sustainable world, but the premonitory signs of a potential catastrophe are now unequivocal, thus urging a structural change in the social and economic organization.

---

96   The biosphere is the thin veil of air, water, rocks and living interactive entities. During the long-term cosmic journey, the biosphere has been subject to several global extinctions, four of which came close to wiping out life on Earth, while several others were local extinctions.

Biophysical, ethical and social limits

The transition from the growthmania — as Daly calls the present stage of the economy — to a steady-state economy via a sustainability process, runs into biophysical and ethical-social limits.

First, the biophysical limits to growth are a result of restrictions imposed by the ecosystem, the finite growth potential of the economic subsystem, and their ecological interdependence. The limits of the hosting ecosystem and its degradation add up, causing the decline of the services rendered to life (see Figure 1.1). As a result, the waste recycling capacity of the environment becomes progressively lower.

Second, ethical and social limits concern the following concepts:

- the growth economy is financed by the depletion of the geological capital (fossil fuels and minerals) and as such is an undesirable cost imposed on future generations

- the degradation of habitats and the decline of biodiversity limit the possibility of continuous growth. Within the natural capital, essential are also the ecological life-support cycles[97] which provide services to plants, animals and people. The deterioration of these services is a further cost imposed on present and future generations

- the unsustainable desirability of a growing affluent society in which every man is willing to become richer than others.

- beyond a certain limit, becoming richer does not mean becoming happier. Daly argues that aggregate growth *"is limited by the corrosive effects on moral standards"* which results in *"the glorification of self-interest and a scientistic-technocratic worldview"*.

---

97   Ecological life-support cycles in the biosphere are long-term established biogeochemical cycles involving the atmosphere, hydrosphere, lithosphere and living entities (plants, animals, microorganisms).

Circular flow, economic and uneconomic growth

According to Daly the transformation from a growth economy to a steady-state economy is inevitable in the long term. This is in line with the view that population growth could most probably stabilize around 2050 or earlier, to be then followed by a decreasing trend. Population and an indefinitely growing economy represent an absurdity in a planet of limited dimension and resources, and the transition to a steady-state economy remains the most desirable mechanism to achieve sustainable development.

In the 1989 book "For the Common Good" by H. Daly and J. Cobb it is suggested that modern economics, should move away from being a *"self-centred academic discipline interested only in working out the consequences of its own assumptions"*[98] and shift from *"Homo economicus as an isolated individual, to the idea of a person in community, whose identity is largely a function of his relationship in community with others and the ecosystem"*.

Figure 7.1 shows the current economy represented as the circular flow in an isolated system, and the transition of the economy from an empty to a full world in which natural resources are limited. The Major inconsistencies of the circular flow is the unrealistic notion of an isolated autonomous self-controlling system, based on the sequence production–consumption–production which ignores the finiteness of resources.

The relationship between an empty and a full world involves the concept that human economy grows within an ecosystem which remains constant in time, thus being a limiting factor of development.

Somewhere between the two extremes of an empty and a full world lies the optimal condition of an ecologically sustainable development, based on resources that are used in quantities no larger than the regeneration capabilities of the ecosystem.

Daly concludes arguing that the first necessity is to stop the exponential growth of the subsystem economy, without falling into the ecological reductionism of an economy totally controlled by the ecosystem.

---

98   Developing Ideas, interview with H. Daly: The Irrationality of Homo Economicus

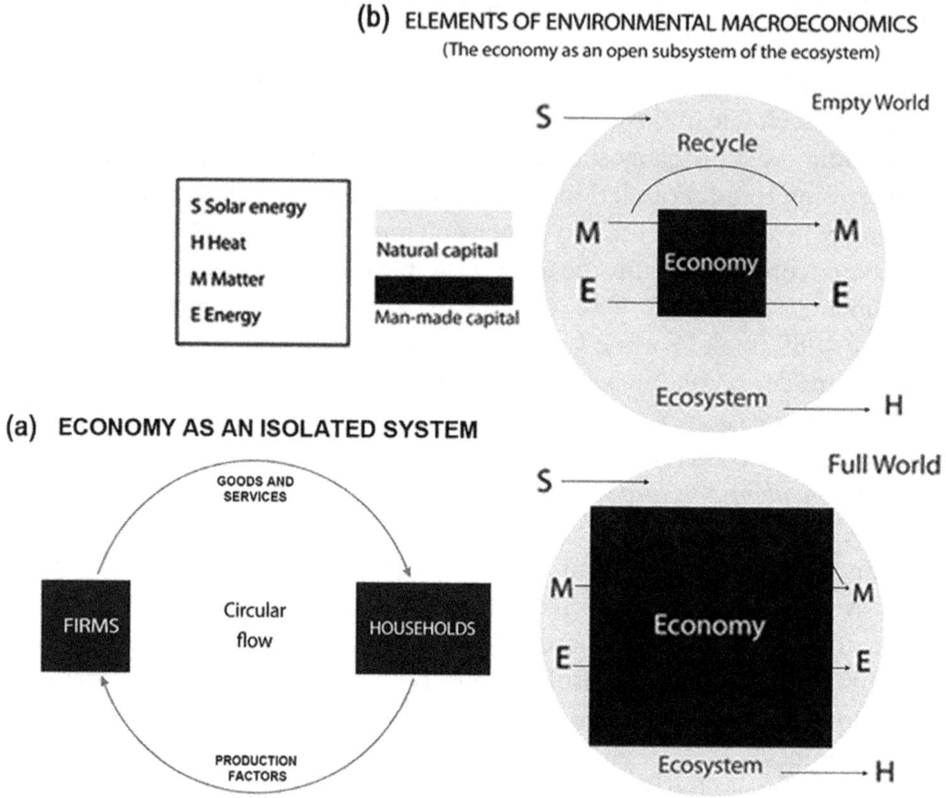

Figure 7.1 – (a) The economy seen as an isolated system. (b) The transition from an empty to a full world. Source: Beyond Growth, 1996, by Herman Daly

The transition from 2.5 billion people in 1950 to a full world of nearly 6 billion in 2000, depicts the change from a small-size economy to an economy grown to the size of the ecosystem. The difference between the two conditions is that the produced capital in 1950s was the limiting factor of economic development, while the present limiting factor is the remaining natural capital.

Figure 7.2 shows the transition from economic to uneconomic growth.

If we imagine the figure to represent the period 1900–2050, the economic limit would have been reached around 1975, the environmental catastrophe would start around 2025 and the "futility limit" be reached by 2040.

ECONOMIC GROWTH AND SUSTAINABILITY

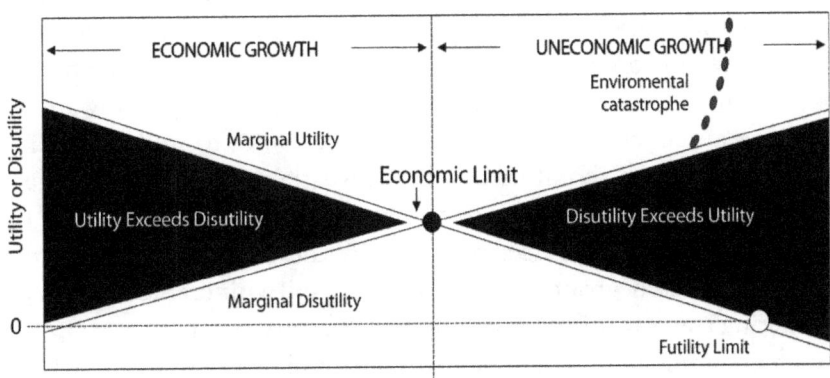

Figure 7.2 – The economic growth, in our ecosystem constrained by limited dimensions and resources, inevitably reaches the economic limit then entering the uneconomic growth domain. The transition point from a decreasing marginal utility into a growing disutility, before reaching the futility limit, triggers the environmental catastrophe. Source: Herman Daly, Scientific American, September 2005.

The original graph by Daly represents a timeless model, in which the economic limit between utility and disutility coincidentally depicts a transition similar in concept to that in the Figure 8.3 from WWF Living Planet Report 2010. Moreover the environmental catastrophe starting around 2025 quite well coincides with the period of catastrophic events from 2020 to 2035 in Figure 9.1 (top), which illustrates the State of the World (2004) by D. Meadows et al. as derived from Figure 8.2. The three models (Daly, WWF and Meadows) are based on different approaches, but their conclusions are similar: humanity has a big problem now and the possibility of an environmental catastrophe is real and close-by.

According to Daly the economic process is based on complementary components: (i) a *natural capital* such as forests, terrestrial and marine life, drinking water and fossil fuels and (ii) the *social (or man-made) capital* which includes all economic activities, basically concerning the transformation and consumption of the natural capital.

The correct balance between these two pillars of our society can be reached through the initiation of an ecologically sustainable process. The natural capital of non renewable resources has been consumed at an accelerated rate in the last decades, drastically reducing the resilience[99] of the biosphere. Renewables are still regenerated by the ecosystem but at a progressively slower rate. This applies to (i) cutting down the rainforest (Amazonia, southeast Asian countries, Africa, America and Asia) which will not be regenerated in decades, (ii) the groundwater pollution by saline water and chemicals, and (iii) the depletion of topsoil, which is regenerated over centuries by the combined effect of weathering and the slow decomposition of organic matter. Humanity is not aware of surviving thanks to long-term natural processes and the services provided by the ecosystem. Deforestation, by leaving an unprotected ground, makes the top soil vulnerable to heavy seasonal rains hampering the possibility of a rapid growth of vegetation.

Growth economists believe that natural capital can be substituted by human-produced capital, ignoring the concept that the two components are not complementary. For an economically sustainable development to take place, human consumption of the natural capital should be downsized and the use of non-renewable resources (fossil fuels and minerals) drastically reduced, giving to future generations a chance to use at least part of them. The environmental crisis is basically a problem of anthropological evolution. The explosive capacity developed in science and technology during the last two centuries was not associated with a wise approach to development. More evident was, instead, an open contrast with an ethics of survival and sustainability!

Gross national product and externalities

GNP[100] and externalities — which are both instrumental to the growth economy — are heavily criticised by Daly. GNP is considered a distorted indicator of the economic progress of nations, since it measures the quantity of production without accounting for negative impacts on the environment and resources depletion. This way it misrepresents the economic progress that it should depict. For example, a country that exports its mineral resources and timber derived from an uncontrolled deforestation (as many African, South American and Asian countries have done for decades), can show a positive GNP growth for years. This growth, however, can be zero or negative if the destruction of the non renewable and renewable capital is accounted for. The formation of a rainforest is a long process in which special conditions of climate, morphology and geology of the surface rocks harmonically cooperate. Clearing the forest can irreparably damage topsoil and

---

99   Resilience is the capacity to recovery.
100 GNP: Gross National Product.

prevent its regeneration. The transition from GNP to SSNNP (Sustainable Social Net National Product), proposed by Daly, involves the economy at the global level. For a correct estimate of production costs, the GNP should be substituted by the SSNNP based on three indexes: services provided, environmental damage (pollution and degradation), and an inventory of the accumulation of materials and funds and their distribution.

This would bring transparency to the production process. However, these three indexes are incompatible with today's method of calculating the GNP. Their global use would require introducing new rules that all nations should accept, which is objectively a difficult task. The change could be started simply by computing on a yearly basis at global and national levels, traditional GNP and the SSNNP for comparison. This initiative could combine well with the activities, duties and responsibilities of the United Nations and be carried out individually by countries as well.

Externalities[101] by the standard economic theory are considered costs or benefits not included in the prices.

In most cases impacts associated with the industrial production are neither charged to producers nor to buyers. Solid, liquid and gaseous waste can alter environment and the ecosystem damage individuals and the community. In 1950s when population was 2.5 billion, externalities affected limited areas in industrialized countries. In the following decades as soon as industrialization developed worldwide, disasters grew in dimension and intensity and people suffered as a consequence of externalities for diseases and a variety of disorders due to pollution.

Economic growth, globalization and Daly's recommendations

The problem of globalization has not been duly taken into consideration by Governments and International Organizations. The market is globally free, but regulations governing the production are different at national levels. Globalization began developing since 1970s when barriers to international trade were progressively lowered. As a result, the World free-trade export (Total World Gross Product)

---

101 Externalities affect economy, environment, ecosystems and the people, thus proving to be most powerful drivers of profit, an obstacle to a fair free market competition and an enormous damage to the society. The costs of medical assistance was partly borne by the public sector, municipalities and more often by the people. The idea of a "polluter pays principle", proposed as a limited solution, proved to be of difficult implementation in the absence rules and control by Government authorities.

doubled between 1970 and 2001 (from 8.5% to 16.2%). According to Daly[102] the globalization should have been achieved through *"the international federation of viable national communities, not through default to a cosmopolitan vacuum left by a world without borders, a vacuum soon filled by transnational corporations"*. Concerning a balanced free-trade he also added:

"Within nations there are laws and institutions that prohibit many cost externalizations. Internationally there are few such laws, and domestic laws and their degree of enforcement vary greatly among nations. Since lower standards mean lower costs and prices, international competition tends to be standard-lowering and thereby destroys community life based on those higher standards. For example, a community whose standards include the avoidance of child labour will not be able to engage in a free trade, with a community that accepts child labour, unless it is willing to lower its standards regarding child labour, or accept the bankruptcy of its business that have to compete with foreign child labour. Either of these alternatives is a severe disruption of its community life".

Five policy recommendations in the economic sector have been proposed by Herman Daly[103], in the speech he delivered in Oslo, while receiving the Sophie Prize 1999 award. The sequence, which goes from the least to the most controversial recommendations, includes:

1. *Stop counting the consumption of natural capital as income*
2. *Shifting the tax base from value added (labour and capital income) onto resources throughput (that to which value is added)*
3. *Maximize the productivity of natural capital in the short run, and invest increasing its supply in the long run*
4. *Move away from the ideology of global economic integration by free trade, free capital mobility, and export-led growth and towards a more nationalist orientation that seeks to develop domestic production for internal markets as the first option, having recourse to international trade only when clearly much more efficient*
5. *Facing the lurking inconsistency.*

---

102   From "Beyond Growth" (1996), page 147.
103   Source: five policy recommendations for a sustainable economy

Conclusive Remarks

The decline of US and EU economies, can destabilize the planet's economy and make more difficult the survival in a time of global environmental distress and growing population. Some analysts are beginning to consider carefully the side effects of globalization and the need for a new global trade order. According to R. Reich[104] consumers' choices have increased, but civil rights and democracy have suffered heavily due to the side effects of the transition to the global market.

Macroeconomy is an open subsystem, constrained by the limits of the ecosystem inside which it develops. The ecosystem in turn provides indispensable services to life and absorbs and recycles waste. As far as the optimum size that the macroeconomy can reach on the Earth, the Plimsoll line (page 91) provides a clear indication concerning physical limits and the optimum equilibrium indispensable for human survival. The initiation of a long-term sustainable development process, through a short-term transition model, can most probably halt population growth in the next 2-3 decades, put climate change under control and help identifying rules to stabilize markets and global trade within a framework of international equity.

From 2050 onward, the decrease in population could lead to a total reshaping of the system by heading towards a steady-state economy within the end of this century. A process of this kind needs to be implemented through the next few decades, if humanity and the ecosystem are to survive decently. Population, the consumption of resources, the ecosystem's protection and climate change are priority issues to be addressed within this decade to make viable the transition from unlimited economic growth to a sustainable development.

In terms of environmental impacts in 2008 China's CO_2 emissions[105] reached $7,031,917 \times 10^3$ Metric Tons per year (MTY), compared to the USA emissions of $5,461,014 \times 10^3$ MTY and EU (27) $4,177,817 \times 10^3$ MTY, respectively representing 23.33%, 18.11% and 14.4% of the global value (29,888,121 MTY). Emissions by USA and EU reach, however, the 32.5% of the total with a population of 800 million, compared to China's population of 1.3 billion and 23.33% emissions.

---

104 R. Reich, "Supercapitalism. The transformation of Business, Democracy and Everyday Life" (2007).

105 Source: en.wikipedia.org

CHAPTER 8

Scenarios for The 21ˢᵗ Century

Global Scenarios

During the past 40 years several attempts have been made to develop possible scenarios in order to understand the future of mankind at a time of transition in which human development is deeply interfering with natural phenomena. Most of these attempts have been short-lived.

Three different scenarios are described:

1. the Millennium Ecosystem Assessment (MA) scenarios, which originated from the initiative of Kofi Annan, UN Secretary-General (2001). The study, based on the review of existing data, was completed in 2005 thanks to the efforts of 1,300 scientists from various nations
2. the scenarios described in a trilogy of books, the first of which was commissioned by the Club of Rome, published in 1972 and authored by a group

of MIT scientists, who continued working on the other two over the next 30 years. The three books — dated 1972, 1992 and 2004 — are based on a system dynamics model covering to the period 1900–2100. The sequence is of great interest, thanks to the updating of data, and the coherence of methodology during the period 1970–2002
3. the scenarios based on the human ecological footprint.

The three approaches focus on major aspects of the global crisis and provide clues to the pathway towards a sustainable planet.

The Millennium Ecosystem Assessment Scenarios

The Millennium Ecosystem Assessment[106] (MA) developed four global scenarios (1900–2100) to explore a plausible future for ecosystems and human well-being, following different global development paths, that is an increasingly globalized world, a growingly regionalized world and different approaches to ecosystems management. Scientists who analyzed the available data reached the conclusion that anthropogenic impacts are heavily affecting the world's biodiversity, making ecosystems and their services to life and humanity decline (Figure 1.1). MA scenarios are:

(i) *Order from Strength*, focusing on security and protection problems, depicts a regionalized and fragmented world, with emphasis on regional markets; it pays little attention to public goods and takes a reactive approach to ecosystem problems. Economic growth rates are the lowest and decrease with time, while population growth is the highest;

(ii) *Adapting Mosaic* is centred on regional watershed-scale ecosystems, which are focal points of political and economic activity. Local institutions are sound, and ecosystem management strategies are based on a strongly pro-active approach. Economic growth rates are low initially but increase with time and the population is nearly as high as in Order from Strength;

(iii) *TechnoGarden* depicts a globally related world based on environmentally sound technologies, using well managed and often engineered ecosystems to deliver

---

106 The project Millennium Ecosystem Assessment (MA) was based on a review of the information available at that time concerning the changes which had affected the ecosystems during the previous decades and the services rendered to humanity. The changes were projected into the future and results were used to construct four plausible scenarios. The MA project warned humanity about the degradation of the ecosystem and negative effects on humanity.

services. The economic growth is relatively high and accelerates, while the population in 2050 is in the mid-range of the scenarios;

(iv) *Global Orchestration* depicts a connected society that focuses on global trade and economic liberalization and takes reactive approaches to ecosystem problems, but also takes steps to reduce poverty and inequality and invests in infrastructure and education. Economic growth is the highest.

Figure 8.1 shows the population forecast for each scenario.

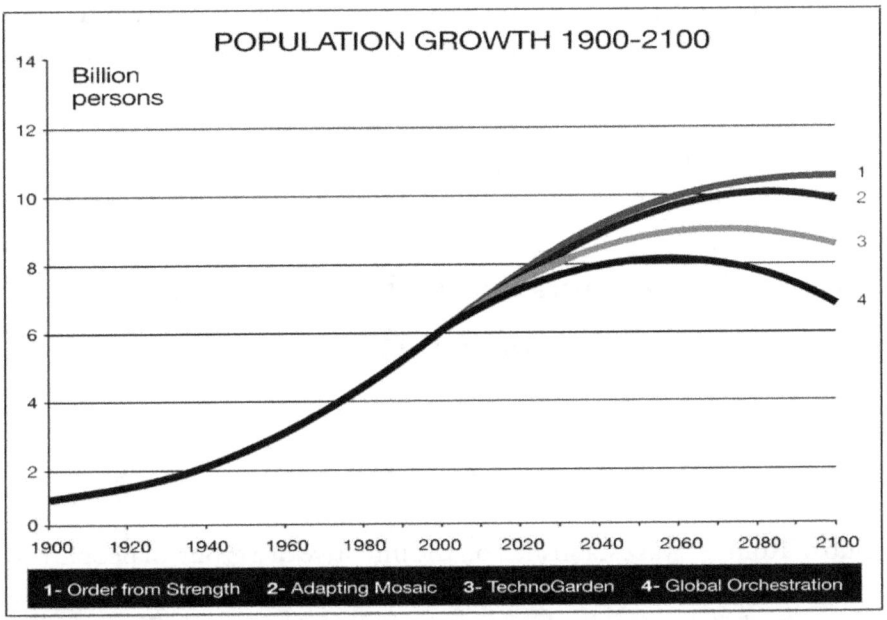

Figure 8.1 – Population growth, 1900–2100. Source: WRI/Millennium Ecosystem Assessment, 2005.

The major findings of the MA Scenarios, developed on the basis of the state of ecosystems and global and regional indicators, are:

- the decline of ecosystems and the associated impacts highlight the need for substantial policy changes and a decisive international action

- human activities are modifying the ecosystem at an unprecedented rate and are thus responsible for the ongoing mass extinction

- if environmental degradation proceeds further neither efforts to eradicate poverty nor the Millennium Development Goals will be achieved.

The MA Report was criticized for:

- the massive text, which was not really useful to policy and decision makers to whom it was addressed

- the fact that the deterioration level of identified ecosystem services[107] — which make the Earth's "ecosystem" hospitable to life — is not assessed

- the proposed four global scenarios appear unspecific, compared to the dimension of problems and potential remedies.

The Scenarios Associated with the Initiative by the Club of Rome

The trilogy and key concepts

The Club of Rome[108] was established on the initiative of a group of people (political leaders, economists, scientists and government officials) who gathered in Rome in 1968, deeply worried about the threats to the future of humanity. Their findings and studies are described in the report *"Limits to Growth"* (1972) which is considered a cornerstone work on global environmental problems. The report paved the way for two further books: *"Beyond the Limits"*, 1992, and *"Limits to Growth". The 30-Year Update"*, published in 2004.

Both reports were also based on the world dynamics mathematical model proposed in 1970 (and updated in the following years) by J. W. Forrester, who had

---

107 See Figure 1.1 (Chapter 1)

108 Aurelio Peccei was the founder of the Club of Rome. In 1968, a group of people from various countries gathered in Rome deeply concerned about the growing threat represented by interacting problems which were becoming more severe with development and which threatened the future of humanity. Their conclusion was that population cannot keep growing at an increasing rate without seriously affecting the natural limits of the system.

already successfully used it to analyse industrial development problems and complex systems.

The first book, *The Limits to Growth* published in 1972, co-authored by four MIT scientists[109], identified the physical limits of the system, foreseeing that economic growth would not continue forever since it would be halted by the limits of available resources. This general approach was adopted and progressively updated during the following decades.

The trilogy depicts the human evolution over 30 years and reports the insights gained over the period from the first premonitory signs of the environmental crisis at the end of the 1960s to the global condition in 2003. The 1994 and 2004 editions:

- utilized the rapid improvements reached in the field of information technology

- took into account historical changes such as the collapse of the Soviet Union and the reshaping of Europe

- revised previous conclusions, included new data on the environment, economic and production/consumption trends and introduced new indicators on the basis of a wider conceptual view.

Each of the three books describes a series of scenarios, including a standard reference scenario considered to represent "high-probability conditions".

Scenarios are based on the hypothesis of variable trends of indicators: availability of non-renewable resources, population growth, food and industrial production rise per capita and global pollution.

Other works were published under the conceptual line adopted by the Club of Rome, but only the *standard reference scenarios*[110] of 1972, 1992 and 2004 trilogy are presented in this Chapter. The period during which the books were published coincided with a progressive degradation of the environment and the ecosystem, the historical period of the "cold war", the collapse of the USSR, the increase in global terrorism and the globalization.

---

109 D.H. Meadows, D.L. Meadows, Jorgen Randers and William W. Behrens III. The books published in 1992 and 2004 on the same topic were authored by the first 3 authors only.

110 Several scenarios are actually described in each of the three books, but the "standard scenarios" are considered to be the most probable.

The trilogy provides an overview of human-related driving factors such as the economy, production, pollution, population growth, the decline of non-renewable resources and biodiversity, and an evaluation of the relationship between environmental deterioration and human interference in the natural domain. The concept on which the models are based is the business as usual (BAU) approach, which indicates a trend rooted in the human lifestyle as it developed in the past.

The expression business as usual, currently used in environmental sciences, was adopted in the 1972 book *The Limits to Growth* to express a general behavioural condition in economy, production and social sciences, invariable in time.

This conservative view in economics supports the traditional actions of buying, producing and selling, inherited from the past. Scenarios today qualified as business as usual indicate the underlying assumption of a stable continuity with the past.

The business as usual approach also implies the conservation of the status quo in economy, a condition which is incompatible with the dynamics of population growth (already beyond the sustainability limit) and the growing demand for food. A scenario qualified as non-business as usual, instead, implies the possibility of new policies to adapt the system to global changes.

Scenarios published between 1972 and 2004

Figure 8.2 illustrates three "standard models", the last two having been updated in the 1992 and 2004 editions based on indications coming from the global environmental crisis in progress.

In the 1972 edition, the collapse of population was foreseen around 2050, while the 1994 and 2004 scenarios placed it around 2035, preceded by the collapse of food and industrial production around 2015. The last version explores in much greater detail the variety of components of the global crisis and its effects. Most politicians and traditional economists do not believe yet that the tipping points of the crisis can be as near as shown in the *State of the World* scenario 2004. The report by the International Monetary Fund (World Economic Outlook) on the 2007–2008 crisis, however, reminded us that old problems remain unsolved and new problems of global economic instability are growing worldwide.

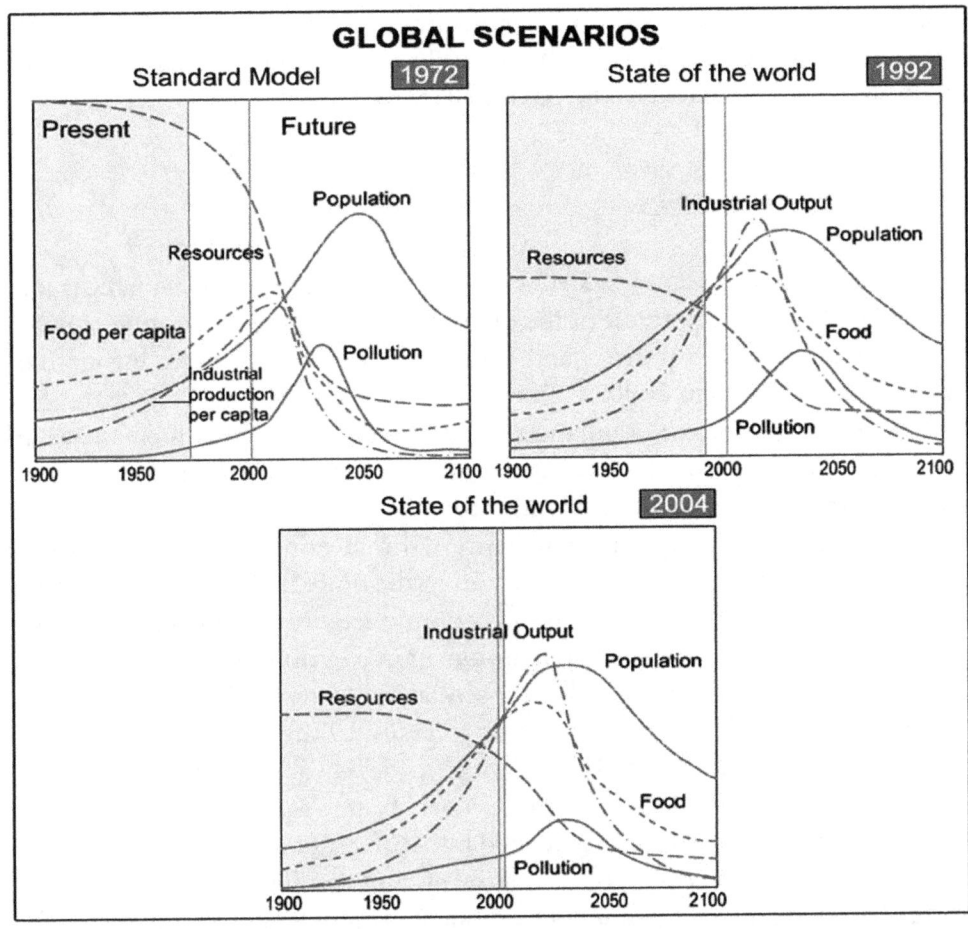

Figure 8.2 – Standard business as usual scenarios for 1900–2100 published in 1972, 1994 and 2004. Sources: 1972 scenario: The Limits to Growth (1972), by Donella H. Meadows et al. Copyright permission by the Sustainability Institute[111]; 1992 and 2004 scenarios from Beyond the Limits (1992), and Limits to Growth. The 30-Year Update (2004,) copyright permission by Chelsea Green Publishing Company.

The physicist Graham Turner, according to the November 2008 media release by CSIRO[112], compared predictions from the 1972 book *The Limits to Growth* to data collected during the following three decades. His conclusion was that the World has continued along the unsustainable trend as a result of the orthodoxy in the *business as usual approach* and traditional views in politics and economics. Critical

---

111 This figure from the Donella Meadows Archive is available for use in research, teaching and private study. For other uses, please contact Sustainability Institute, 3 Linden Road, Hartland, VT 05048, (802) 436-1277, USA.

112 Commonwealth Scientific and Industrial Research Organization (Australia)

tipping points of indicators show the tendency to anticipate: in 1972 these were centred on the 2010–2050 period, but in 2004 the concentration of critical points of indicators was shortened to the period 2015–2035.

Concluding remarks

The forecast of *The Limits to Growth* (1972) concerned the unsustainable trend of the growth economy as a result of the orthodoxy in the *business as usual approach*, the traditional views in politics and economics and the accurate estimate of a population of 6 billion in 2000.

The general picture of population rise and industrial capital increase, as drivers of the economic growth, was further analyzed in the 1994 and 2004 books, together with a growing number of indicators.

Throughout the trilogy the view is stressed that non-renewable resources, on which the economy depends, are constrained by the finite limits that apply in general to the physical world and by impacts of economic activities on the natural environment. The declining capacity of the environment in recycling waste and the dramatic effects of climate change are considered essential indicators in the last two books.

The initial criticism to *The limits to growth* (1972) concerned the use of a dynamic model (by J. Forrester, *World Dynamics*, 1971) based on statements, observations and assumptions derived from the world system as it is: interconnected and interacting subsystems, each characterized by a different behaviour.

The debate concerned also the accusation of "catastrophism", the insufficient basis of data concerning the variables and the exponential growth of some indicators. Today the exponential trend of indicators is no longer questioned and predictions provided by the trilogy are considered by a growing number of scientists to be basically right. It is worth recalling A. Peccei's conclusion in his 1971 book *"Which future?"*:

- on the basis the dynamics of the economy and the business as usual approach, a final crisis will most probably be preceded by wars and the return of human society to a tribal condition

- a number of strong nations could control the World and attempt to block the collapse by imposing a global military dictatorship inevitably increasing differences between rich and poor

- or, eventually, the best human qualities could prevail, and reason could hinder dictatorial solutions.

The Ecological Footprint Analysis

Human footprint, biocapacity, overshooting and ecological debt

In the early 1970s William Rees developed the concept of "a regional capsule" — later on the Ecological Footprint — to stimulate students involved in multi-disciplinary planning to develop the debate about human demand on ecosystems and the Earth's carrying capacity.

The 1996 book *"Our Ecological Footprint, Reducing the Human Impact on the Earth"*, by M. Wackernagel and W. Rees, describes the methodology to reduce the human imprint on the planet, recognizing that the global crisis primarily implies behavioural and social aspects and at a lesser extent environmental and technical problems.

The ecological footprint analysis (EFA) compares, on a yearly basis and for a defined population (individual, national and global levels), humanity's demand with the capacity of the ecosystem to renew resources and absorb waste. EFA assumes that resources, the quantity of food and the services rendered by the ecosystem derive from nature and are therefore limited. The main indicators are (i) the ecological footprint as the measure of the human demand on Earth and (ii) the bio-capacity as the ecosystem's ability to reproduce biological material and absorb waste. When the human footprint grows beyond the Earth's bio-capacity, ecological overshooting takes place and the ecological debt increases. The current consumption of resources grows faster than the natural regeneration capacity of the planet.

The ecological footprint — expressed as the area (hectares) of land and sea needed to provide human society with resources and the waste absorption zones — is conditioned by the growth of population, the rising lifestyle of people, the restless increase of human activities and impacts and the huge differences among nations. Biocapacity — also measured as an area (hectares) — is defined as the capacity of a biologically productive area to generate renewable resources and absorb waste.

The comparison of the two indicators for a given bio-capacity of the planet, shows the number of Earths needed to support humanity. The bio-capacity of the Earth[113]

---

[113] In 2003 the surface of the Earth was evaluated in 51 billion hectares, 36 billion of which were occupied by oceans and seas, rivers and lakes and 15 billion by land areas not covered by ice. Out of these 15 billion, 0.8 billion were unproductive, while the other 14.2 billion represented the 2003 global bio-capacity of the Earth. Dividing 14.2 billion hectares of productive land by the population of 6.3 billion people in 2003, the average bio-capacity per person was 2.2 ha.

is represented in billion hectares with a yearly based reproductive potential. As long as the human footprint does not exceed the Earth's bio-capacity the system is sustainable.

Figure 8.3, concerning the period 1961–2007, shows (a) the global ecological footprint by components and the accumulated ecological debt, and (b) variations of the ecological footprint and biocapacity per person.

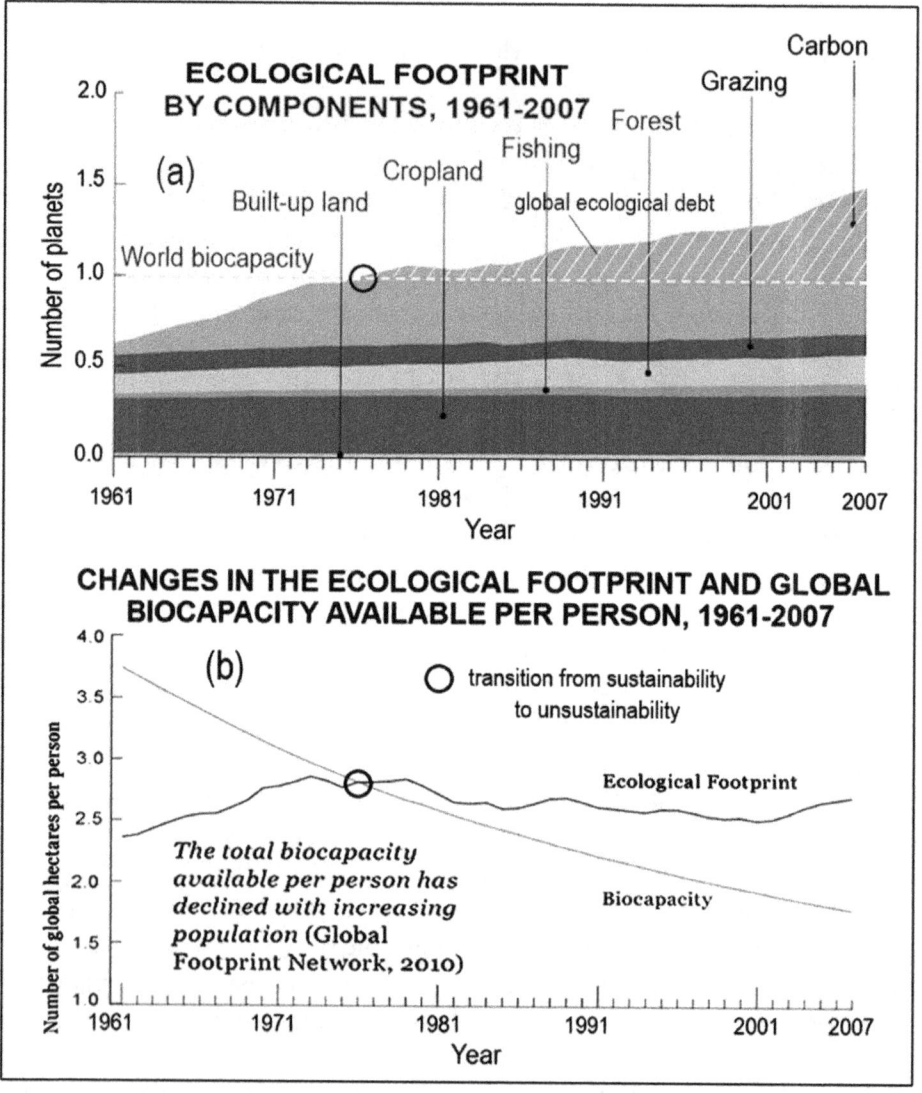

Figure 8.3 – (a) In 2007 humanity reached the level of 1.5 planets (global debt accumulation), consuming 50% more resources than Earth could renew in that year; (b) during the period 1961-2007 (when the population increased from 3 billion in 1961, to 6.7 billion in 2007), ecological footprint and biocapacity curves crossed one another in 1976. Source: WWF Living Planet Report 2010.

In the mid 1970s the ecological footprint reached the world bio-capacity limit of one Earth, then humanity started accumulating ecological debt and entered the unsustainability domain which inevitably leads to collapse. Since 1976 humanity has used more resources than nature has been able to renew and the process brought to the accumulated global unsustainability level of 1.5 Earths in 2007.

Half the world biocapacity today is shared by 10 countries (in decreasing order Brazil, China, USA, the Russian Federation, India, Canada, Australia, Indonesia, Argentina and France). This unequal distribution depicts the huge diversity and variety of bio-capacities worldwide and the inevitable imbalance between the richest ten countries and the others. The consequent rise in human demand on the biosphere generates a growing global impact to which the deterioration of soil, water and ecosystem services is associated. Plato[114], more than 2 millennia ago expressed the concept that the number of citizens had to be compatible with a land extension that could support them comfortably.

A similar concept of stability and safety is used in the design of bridges (and other structures), based on the accurate evaluation of the limits of the system, the bearing capacity of the structure, the static weight involved, dynamic loads (vibrations induced by vehicles), the speed and direction of winds and so on. Bridges and structures in general are systematically repaired through routine and special maintenance to preserve their original carrying capacity.

Should the same concepts be applied to the biosphere, then human imprint and biocapacity, the recognition of the ecosystem's limits, the carrying capacity of the biosphere and, above all, the growing population should become central planning indicators for a common survival. The World our ancestors inherited at the time of the farming revolution was practically unlimited in terms of biocapacity, and the human footprint inevitably minimal due to the scarce population.

The growth of human activities and impacts started rising exponentially with the industrial revolution in 1800s and then, in mid-1970s, humanity reached the limit at which footprint equalled biocapacity and the people unknowingly entered the unsustainability domain.

Moreover ecological conditions, biodiversity, the density of population and the ecological debt are unequally distributed, thus making the system growingly vulnerable.

---

114   Plato's view about human carrying capacity is cited (page 48) in the 1996 book "Our Ecological Footprint", by M. Wackernagel & W. Rees.

Ecological footprints, biocapacities and freshwater consumption

Figure 8.4 shows the 2007 ecological footprint per country, per person.

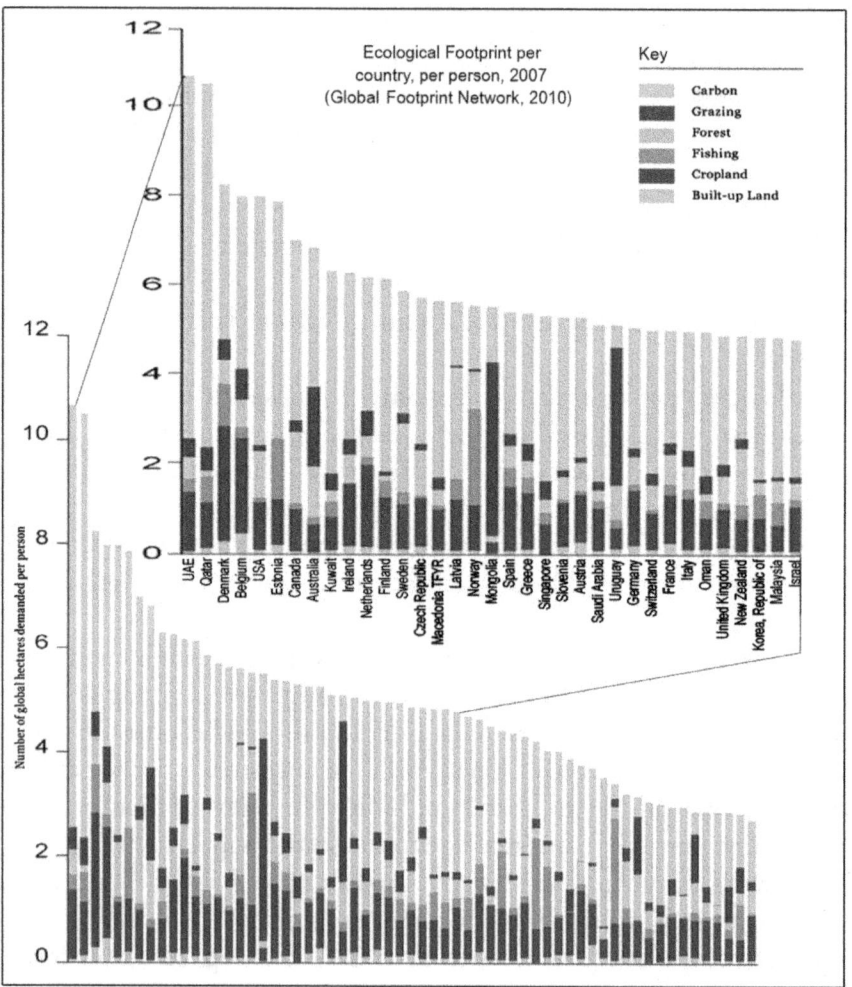

Figure 8.4 – The Ecological footprint per country per person is based on the population of each country in 2007. The original figure includes 164 countries, each histogram being based on six components: Carbon, Grazing Land, Forest Land, Fishing Grounds, Cropland and Built-up Land. The bottom graph shows the footprint trend of the first 61 countries, the top graph illustrating the enlarged window of the first 35. The imprint of carbon largely prevails on other key components. Adapted from WWF Living Planet Report 2010.

Figure 8.5 shows the biocapacity per person per country in 2007.

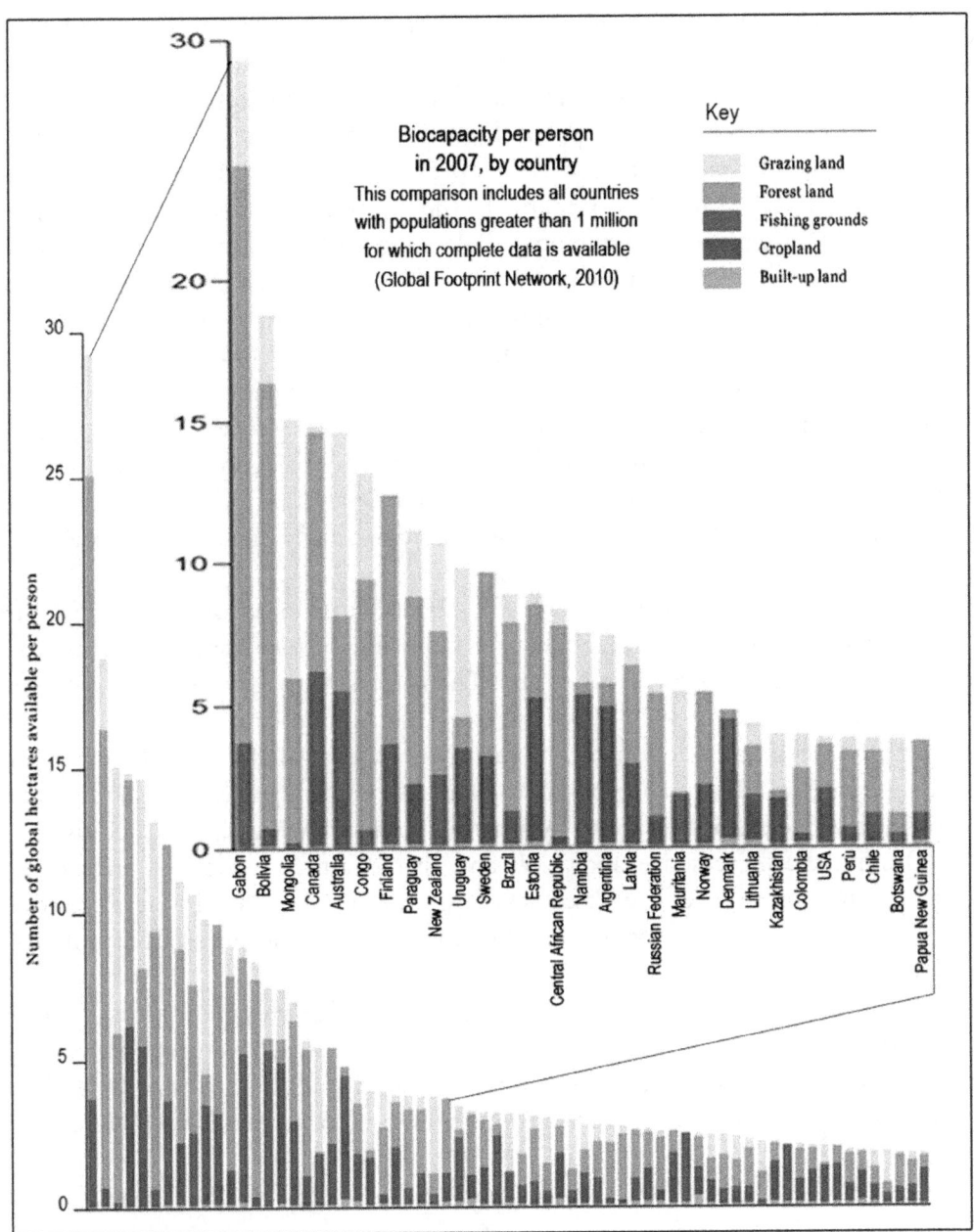

Figure 8.5 – Biocapacity trend per person per country in 2007. Out of the 152 countries included in the original graph, only half are represented in this Figure (bottom), the enlarged window (top) concerning the first 25 countries only. Adapted from WWF Living Planet Report 2010.

Biocapacity is based on four types of productive land areas[115], plus fishing grounds and is also measured by the number of global hectares available per person. The conclusion is that neither ecological footprints nor bio-capacities of nations are evenly distributed, the global picture resulting unequivocal: the overshooting level and debt accumulation impose an unbearable cost to present and future generations.

The footprint-biocapacity approach carries some limits in the definition of terms. Histograms of individual nations provide a general picture of the excessive imprint of rich countries compared to the limited consumption of developing countries which represent the majority. Balancing the gap requires a sustainability process which is hard to imagine in a society obsessed by the GNP. New technologies for clean and cheap energy, a sustainable development process, an improved cultural level of the people, a decreasing population growth-rate and cooperation among nations appear indispensable, but reaching the balance between the ecological footprint and bio-capacity remains a long-term process. The Ecological Footprint Analysis (EFA) can be considered a cognitive basic tool for a preliminary evaluation of the human imprint and overshooting and an indication to evaluate the sustainability gap between ecological production and human consumption and the way it can be reduced.

The global water withdrawal in 1900–2000 and projection to 2025 (Figure 8.6), provide an indication of global water use in agriculture, industry, municipalities and the quantity stored in reservoirs. The increasing irrigation accounts for the largest water consumption in 2003, compared to the moderate increase by industry and municipalities. A controlling factor of water availability is deforestation, which reduces rainwater absorption and fastens the flow to the sea. Aquifers are also affected in developed nations by growing pollution which limits their use.

Despite the increased evaporation and rising rainfall associated with global warming, in some areas of the planet less freshwater is available, in others the growing quantity and intensity of precipitations is causing floods and faster and more concentrated discharges.

A rational consumption of water for a sustainable development process should take into consideration a variety of factors, including losses in water supply systems, more efficient irrigation methods, the reduction in polluting substances and the damaging effect of human impacts among which deforestation and pollution are the worst.

---

115   Types of productive land areas are: (i) livestock and non-food crops, uncultivated land for the production of meat, skins, wool and honey, (ii) forest land for the production of wood, fuel, fibre and paper, and forests that assimilate and sequester carbon dioxide, (iii) agricultural land for food, and (iv) land for infrastructure, building, transport and industrial production.

Global warming alone is considered a significant driver of freshwater decline, by the accelerated melting of ice covers worldwide, which in turn implies the rapid discharge of freshwater and more frequent floods in low-lying areas. The Global water consumption is shown in Figure 8.6.

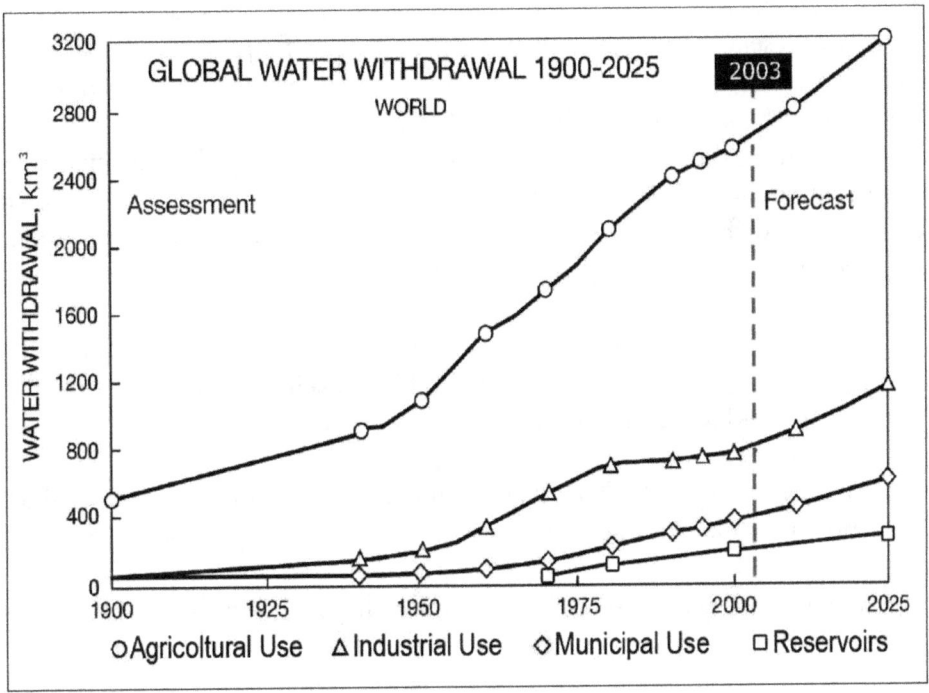

Figure 8.6 – Global water consumption and forecast to 2025. The rapid increase in consumption started around 1950 reaching in 2003 a volume 8 times larger than in 1900. The consumption in 2025 is expected to reach 5,200 km³ (3200 km³ agriculture, 1200 km³ industrial use, 600 km³ municipal use and 200 km³ reservoirs), a value 10 times larger than in 1900. Source: Igor A. Shiklomanov, State Hydrological Institute (SHI), St. Petersburg and UNESCO, Paris 1999.

The thinning of the ice mantle in the Tibetan Plateau and along the Himalaya Range puts a population of nearly 2 billion people in Asia at risk. Every year nearly 110,000 km³ of rain falls on the planet's surface. About 70,000 km³ are used by vegetation and the remaining 40,000 km³, discharged by the global river network, represent the quantity of fresh water annually entering lakes, rivers and aquifers.

Out of the 40,000 km³ an estimated demand of 5,200 km³ is expected by 2025 for human use, depending on climate conditions (the local rainfall, the distribution over the year and the rate of evaporation) and the water quality of aquifers.

Water scarcity is already a problem in Sahel and other dry land areas worldwide. At present 45 nations — over 130 — according to the *WWF Living Planet Report 2010* — experience moderate to severe water scarcity. The current human consumption of water is central to the production/consumption system, which ignores the need to preserve forests to enhance the recharge of aquifers and prevent their pollution. By 2025 a rising water demand is expected for agricultural and industrial use to support the needs of 8 billion people.

By combining data concerning the ecological footprint, biocapacity, scarcity of water and unequal social development, a highly diversified picture of humanity emerges; rising difficulties are expected, due to freshwater scarcity in some areas and related problems for food production. Main factors affecting water scarcity are:

- population growth, increasing urbanization and industrialization will definitely generate a growing demand in areas in which water is already limited and locally polluted.

- climate change which is likely to make the distribution of precipitations more irregular in time and location.

In sum human expansion and freshwater availability follow different pathways.

Implications of current human development and scenarios

Major consequences of the 1970–2010 economic development are:

- *climate change* — induced by the current emissions of GHGs — which threatens the whole ecosystem, heavily degrading arctic and alpine environments, increasing the melting of land and sea ice, causing coral bleaching, altering natural cycles, inducing droughts, desertification, biodiversity decline, habitat loss and the spread of invasive species

- the *consumption of biological capital*. The growing demand for food requires a greater utilization of fertilizers, insecticides and herbicides, which increase the pollution of aquifers and other water bodies (rivers, seas and oceans) and lower the fertility of the soil;

- the *distribution of photosynthetic production*. In 1986, Peter Vitousek, a biologist at Stanford University, estimated that in the 1980s, 40% of global biological production from photosynthesis was absorbed by human activities, decreasing the share for other species (some threatened with extinction). Fish production (marine and inland water wild fish catches) reached its maximum of 88 million tonnes in 1990, then remaining stable around 90 million tonnes between 1990 and 2006 (FAO, Wild Fish Statistics, 1950–2006)

- the *economic inequality*, which concerns the uneven distribution of property and income, is a combined effect of human activities, historical trends and the economy based on the "business as usual" approach. In 2007 around 1 billion people shared 60% of the global property and income.

Figure 8.7 represents scenarios from WWF Living Planet Report 2006.

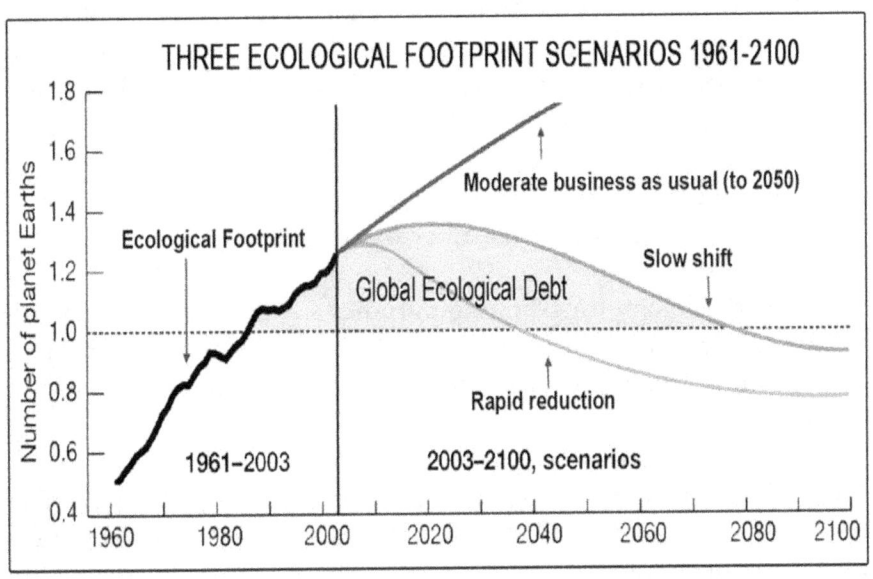

Figure 8.7 – Human ecological footprint scenarios between 1960 and 2100. Three possibilities are considered: humanity continues with a moderate business as usual approach, reaching the level of 1.8 Earths in the 2040s, or accepts the sustainability process for the reduction in the accumulated ecological debt by following a slow shift trend. The rapid reduction of the ecological debt from 2010 is evidently over. Source: WWF Living Planet Report 2006.

Only the "slow shift" scenario, with a declining ecological footprint from 2020, can fit today with a sustainable development pathway.

Pros and cons of the ecological footprint analysis (EFA)

Criticism and praise for the ecological footprint, bio-capacity, ecological debt and carrying capacity concepts were expressed during the past decade. Major concerns involved the definition of terms, the methods of measurement, the components and finally the usefulness of the method as a sustainability assessment tool. Findings of the EFA are:

- the global crisis is basically attributable to human development, rather than to the biosphere's dynamics. Footprint and biocapacity (similarly to other aggregate indicators), are based on global hectares which normalize into an area different components, like forests, fishing grounds and carbon emissions, by using equivalence conversion factors based on productivity. The current footprint by trespassing the limit beyond which biocapacity declines, has made clear that our world is growingly unsafe and unsustainable;

- the concept that in 2007 the ecological imprint reached 1.5 Earths (Figure 8.3 a) is a dramatic self-explanatory warning;

- as a methodology for planning purposes EFA, represents a suitable tool for the inventory of resources available and human consumption, needing however further improvement to turn it into a planning process suitable for policy-makers.

As stated in the book "Our Ecological Footprint" (1996) by M. Wackernagel and W. Rees, EFA is not a "predictive tool", rather an ecological camera of current human demands on nature.

Remarks on EFA concern:

- the definition of terms and assumptions, which in some cases appear contradictory, and the conversion of indicators into hectares as a measure of footprint and bio-capacity.

- technological, cultural and scientific developments are not considered.

- the carbon footprint appears overestimated, while the decline of bio-capacity seems underestimated.

It is time now that nations worldwide start implementing a sustainability process by maximizing inside their borders the balance between human footprint and the renewal of resources and by strictly controlling the global debt accumulation which can lead to bankruptcy.

The Ecological Economics Approach

Even though it does not concern scenarios, it is worth recalling the scientifically sound analysis for an interdisciplinary planning approach proposed in the book *"Ecological Economics, A work-book for problem-based learning"*, (2005) by J. Farley, J. D. Erickson and H. A. Daly. Authors illustrate a goal oriented process to solve problems through ecological economics, providing examples and local solutions. Although described examples concern specific problems, guidelines can be used to solve greater problems to change the world. The problem-based-learning recognises the diversity of local conditions, the need for tailor-made solutions and the possibility for problem-solvers to identify procedures that can be applied to the variety of conditions. Problem-based learning helps in solving international, national and a variety of local problems by following four steps: problem identification and definition, analysis, synthesis and communication of results. The procedure is based on D. Meadows'[116] nine leverage points for forcing a system to change:

- playing with numbers when the system is close to the threshold

- material stocks and flows

- regulating negative feedback loops

- driving positive feedback loops

---

116   Dr. Donella Meadows (1941-2001), Ph. D. in Biophysics, Harvard University, was the author or co-author of nine books among which the trilogy of books Limits to Growth (1972), Beyond the Limits (1992) and Limits to Growth. The 30-year update (2004).

- information flows
- the rules of the system
- the power of self-organization
- the goals of the system
- the mindset or paradigm.

Ecological Economics, in contrast with neoclassical economy based on the business as usual, is a multidisciplinary approach deeply rooted on social and natural sciences and an ethical background as a support system for a sustainable development process.

CHAPTER 9

The Sustainability Development Process and The Reference Transition Model

Humanity in a Phase of Transition

The current global environmental crisis is an unprecedented event in the geological history of the planet and is a result of the interaction between man's current activities (and impacts) and natural long-term phenomena. The present emergency stems from these two components, the former of which brought the system near the current limiting equilibrium condition.

The transitions towards a sustainable development was proposed in 1987 by the Brundtland Report (United Nations World Commission on Environment and Development), as a solution *"which meets the needs of the present without compromising the ability of future generations to meet their own needs"*.

Unlike three decades ago, today a great amount of data has been gathered on the current critical condition of the biosphere and new policies and technologies have been identified for the initiation of a sustainable development process. Solutions, however, are not simultaneously adoptable by nations, since their cultural, social, economic and behavioural diversities at the moment hamper the achievement of international agreements. Initiating a process of change, remains the most urgent action that only large countries or groups of countries can unilaterally adopt, other nations hopefully joining later. Several issues are met on the pathway towards sustainability:

- exposure and vulnerability, changing across space and time scales, depend on a variety of factors (population, economy, social and cultural structures, environment, human impacts and governance)

- the decline of non renewable resources

- the uniqueness of life within the solar system. The special tuning of natural phenomena[117] and the balance of forces, which allowed events to unfold during a long-term evolution into a progressively hospitable planet

- the complexity of human development and associated impacts which are today poorly understood by the majority of people

- the time gap between the accelerated trend of the current global crisis and the slowness of human reaction. Advances in science and technologies have greatly enhanced human progress, but not accelerated the evolution towards the higher maturity level needed to overcome the emergency. Humanity is confronted today with the problem of desiring an affluent future based on the business-as-usual approach, which in turn drives the decline of ecosystems and associated services indispensable for a stable life.

---

117 The daily quantity of sun radiation absorbed by the Earth, the natural greenhouse effect which makes the biosphere just warm enough to host life, the protection provided by the magnetic shield (which prevents the Earth's atmosphere from being wiped out by solar wind), and the ozone layer are phenomena which evolved in a life-friendly trend. The history of the Earth, however, was also characterized by abrupt changes, which repeatedly threatened life. One of these accidental and sudden events could turn to be human development.

Human activities and impacts, as they are now, threaten the stability of the entire biosphere

- the knowledge gap between scientists on the one side and politicians, economists and part of the of the population on the other, represents a formidable difficulty in sharing global solutions.

Unlike other species, always defenceless during past climate changes[118], humanity can take advantage today of science and technologies. This implies, however, accepting the concept that natural changes on Earth were innumerable in the past, that we live in a vulnerable place and that life and the environment are now in peril. This new awareness is still uncommon in our society and difficulties to achieve it are many. Life in the past has been rooted in the stability of the surrounding environment (climate, vegetation, natural resources), and the competition for survival, which has never been purposely colliding with natural life-friendly processes. Human survival and related economic activities, instead, have perturbed natural cycles and ecosystem services indispensable to the biological world. Human collision with the nature's capacity to renew resources destabilizes the ecosystem by isolating its component subsystems and altering their natural behaviour. The Special Report by the IPCC "Managing the Risk of Extreme Events and Disasters to Advance Climate Change Adaptation" (SREX), was announced to be available in February 2012. The Summary[119] for Policy Makers (December 2011), describes "extreme events" and effects associated with exposure of people and the vulnerability of infrastructure and social and economic assets. The need for disaster risk management and strategies is stressed in the light that extreme events are likely to increase during the next decades.

The Nine Major Indicators of the Crisis

Today nine environmental phenomena[120] are close to or beyond threshold limits, drastically reducing the capacity of the Earth to maintain the stability condition indispensable to humanity.

---

118 About 20 ky ago started the last deglaciation, followed by the rise of surface temperature. Around 14 Ky ago a sudden cooling stage named Dryas occurred. Starting 10 Ky years ago climate remained nearly stable until 1800, when global warming began.
119 Ipcc-wg2.gov/SREX/report
120 Source: Tipping towards the unknown. The group of scientists headed by J. Rockström of the Stockholm Resilience Centre (Sweden) identified a "safe planetary operating space" in which humanity could continue to develop for generations.

The nine phenomena which were identified by the scientists of the Stockholm Resilience Centre are: climate change, stratospheric ozone depletion, land use change, freshwater use, biological diversity decrease, ocean acidification, nitrogen and phosphorus accumulation into the biosphere and the oceans, aerosol loading and chemical pollution. For each of these phenomena, pre-industrial values, current values and limits have been identified. According to the study:

- three of these interconnected processes — climate change, biodiversity, nitrogen and phosphorous flow into the biosphere — have already passed the limit, but remaining close to the "tipping points"

- four — land and freshwater pollution, stratospheric ozone depletion and ocean acidification— are irreversible already"

- the last two — aerosol loading and chemical pollution — are still being investigated for the identification of values.

The evaluation is based on different parameters[121]: for example in the case of climate change the parameter is CO_2 ppm concentration, for freshwater it is the consumption in cubic kilometres per year and for land use it is the percentage of land converted to crops.

Signs of a climate transition are many and among them is worth recalling the melting of land glaciers worldwide, the decline of Arctic ice sheet, and the retreat ice and snow covers along the Himalayas, the Tibetan Plateau and the Andes, Greenland and Antarctica.

Climate change, biodiversity loss and the widespread pollution show that our planet is near the upper limit of the *Anthropocene*[122], which can end with a global catastrophe.

According to the article *Target Atmospheric CO_2: Where Should Humanity Aim?* by James Hansen *et al.*[123]:

- the present level of 385 ppm CO_2 is considered far too high to preserve the long-term climate stability during which humanity developed

---

121   Source: Scientific American, April 2010, Boundaries for a Healthy Planet, by J. Foley.
122   Anthropocene is the term coined to define the period – from the industrial revolution in the 1800s to the present – dominated by human history
123   James Hansen, Head of the Goddard Institute for Space Studies (NASA) in New York.

- a reduction in atmospheric CO_2 to 350 ppm (compared to 390 ppm in 2010), is urgently needed to halt the worsening of the climate

- the inevitable effects of CO_2 include ocean acidification, freshwater loss, the migration of climatic zones and impacts on ecosystems

- further growth of GHG emissions for another decade is likely to hinder the possibility of stabilizing the atmospheric composition below the tipping level at which catastrophic events can occur

- *"the most difficult task, phasing out over the next 20–25 years coal use that does not capture CO_2, is herculean, yet feasible when compared with the efforts that went into World War II. The stakes for human life on the planet surpass those of any previous crisis. The greatest danger is continued ignorance and denial, which could make tragic consequences unavoidable".*

The State of the World

Figure 9.1 shows, (top) the State of the World 2004 by D. Meadows *et al.*, (from Figure 8.2), (middle) the trend of major driving indicators and, (bottom) the proposed action, which includes a long-term sustainable development process (SPD 2010–2080) and a short-middle term action, based on a reference transition model (RTM 2010-2040).

The State of the World 2004, rooted on the business as usual approach, depicts the possibility for our society to undergo a critical sequence of global disasters between 2015 and 2040. The closest tipping points, concerning food and industrial outputs, however, are likely to become critical in 2020 or so. Therefore, on the basis of the 2007–2011 economic crisis and recession, humanity appears already involved in a global emergency!

The middle graph illustrates the topmost drivers: (i) a rising CO_2 concentration cause of global warming, and, (ii) the population which by the end of this decade will almost certainly reach 7.5 billion people, with dramatic effects on the global market system. In sum carbon dioxide emissions and population growth, both unstoppable by 2020, will determine the trend of other dependent indicators, such as food availability, the decline of non-renewable resources and the industrial output. The continuous increase of population, despite the current decline of the annual growth rate, remains a most dangerous indicator.

THE GLOBAL ENVIRONMENTAL CRISIS

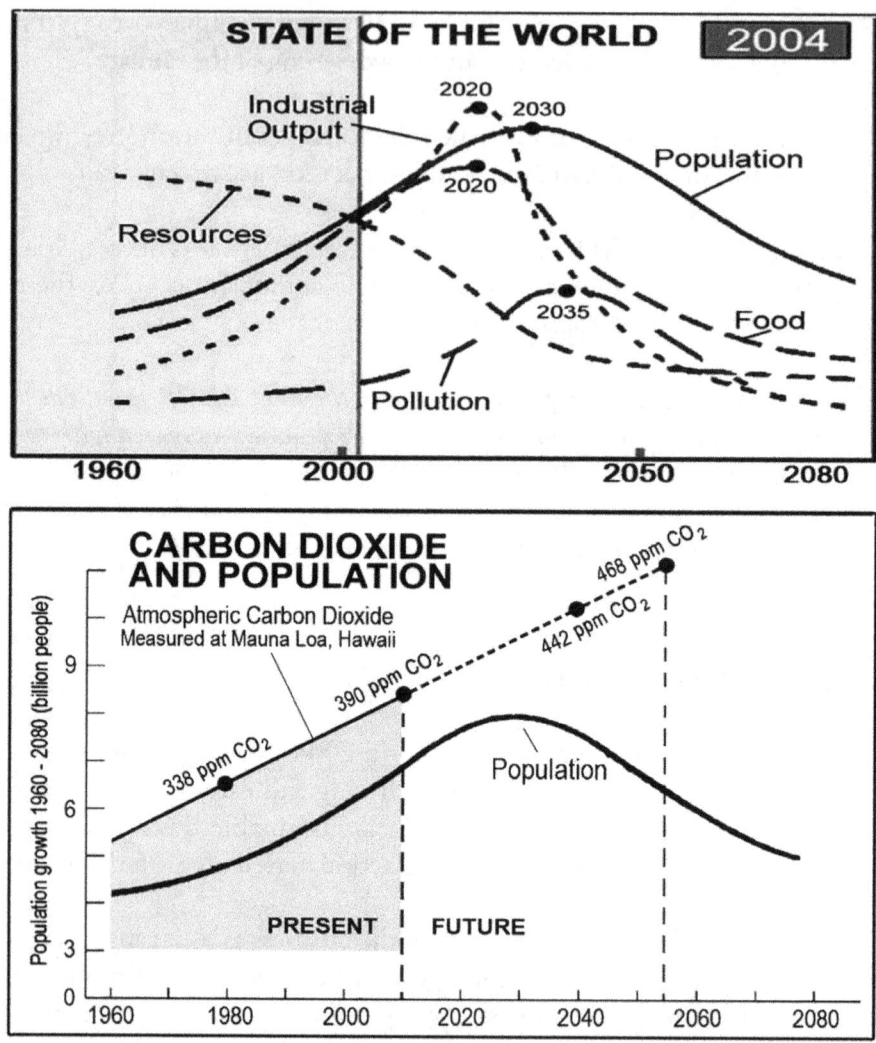

Figure 9.1 – (top) The State of the World 2004, – based on resources, food, industrial output, population and pollution – reaches the critical condition between 2020 and 2035. Sources: (top) Limits to Growth. The 30-Year Update *by D. Meadows et al., used under the permission of Chelsea Green Publishing, www.chelseagreen.com; (middle) CO_2 emission increase from scrippsco2.ucsd.edu and population from the box above; (bottom) Timelines for the Sustainable Development Process (SDP) 2010–2080 and the Reference Transition Model (RTM) 2010–2040.*

The declining food production in the coming years or the rising prices or both, can drastically increase the number of undernourished people (around one billion in 2011) and destabilize the global economy.

Carbon dioxide emissions — as already warned by J. Hansen — are out of control. Data available by the end of 2011 show that the meltdown of permafrost in the northern hemisphere is growing fast, making one of the biggest global "sinks" of carbon to contribute significantly to global warming, through the release of unprecedented quantities of methane.

With this in mind I have concluded from Figure 9.1 that humanity should attempt a new pathway by adopting

1. a **Sustainable Development Process** (SDP) 2010–2080 in the hope that before the end of this century an ecologically sustainable condition is reached under the continuous effort of the United Nations and a globally shared participation of countries
2. a **Reference Transition Model** (RTM) 2010–2040, to overcome the emergency, which will probably reach its most critical phase within the next decade. RTM should be adopted and carried out individually by big nations or by groups of nations, before the sequence of uncontrollable events begins. Primary relevant indicators are population and GHGs emissions.

The State of the World 2004 scenario is not considered by the authors of *"Limits to Growth. The 30-Year Update"* (D. Meadows *et al*). as a prediction in terms of dimension and time, rather the most probable future of our World, if the economy remains strictly based on the business as usual approach.

The abrupt rising of food prices[124], which took place in 2007 and remained critical during the past four years, was accompanied by the massive migration of people from Africa to Europe. According to UNICEF in 2011 a widespread drought is still affecting Eastern African countries, heavily involving about 20 million people half of which in Ethiopia. The trend of food prices is mainly the result of the fast growing demand by emerging economies and rich countries.

Humanity's condition can be represented as an emergency conceptually similar to the situation of the Titanic: the floating iceberg has been identified but the ship seems neither in a condition to change direction in time to avoid the obstacle, nor to minimize the impact. At the moment (December 2011) the global economic

---

124 FAO Food Price Index started increasing moderately around 2000, abruptly rising in 2007, reaching a top level in 2008 and then falling down in 2009 to 2007 level. In 2010 a new sharp rise was resumed and is still ongoing in 2011.
Source: 2007–2008 world food price crisis (Wikipedia)

crisis is on a most critical stage and, if global warming proceeds at the current rate, humanity will face further deterioration of the ecosystem, growing migration of invasive species, drought, higher intensity rainfalls and flooding.

Sustainable Development and the United Nations

Since the 1972 Stockholm Conference on the Human Environment, the United Nations concentrated on various international issues. Hence, the SDP is one of their primary tasks in keeping open the debate on sustainability by emphasizing risks and trends and dealing with urgent problems. The UN long-term action, even when difficult and partially inconclusive, represents the main pathway to the future. Fundamental cornerstones of the UN are

   i. the 1987 Brundtland Report which highlighted the possibility, the usefulness and the urgency for a new era of sustainable human progress, based on policies that can support the environmental resource base and reduce poverty in the developing world;
  ii. Agenda 21 (which is the action plan of the 1992 Rio Conference on Environment and Development) and the subsequent Summits which confirmed the UN commitment to the achievement of Millennium Development Goals
 iii. International Conferences organized by the UN during the last decades.

The SDP 2010–2080, under the responsibility of United Nations, could turn to be a major 21st century task. A stronger effort by the UN, within their wide spectrum of initiatives, should concern:

- deforestation, overfishing, excess in exporting wood fibre and fossil fuels, accelerate the decline of very basic resources, which are all affected by overconsumption; this impoverishes the capital of nations, but allows local governments to enjoy an apparently growing GNP. Halting deforestation helps storing more carbon dioxide, fostering the photosynthesis, protecting biodiversity and preserving ecosystem services. Vegetation[125] is essential in stabilizing climate, in

---

125 It is worth recalling that vegetation transforms solar energy into carbohydrates. In this view leaves play the role of thin natural "solar panels", in the absence of which the sun's radiation could not be captured and transformed.

providing a suitable habitat for innumerable species worldwide, in transforming solar energy into carbohydrates and providing irreplaceable services

- the change of the GNP into the new indicator SSNNP as suggested by Herman Daly (Chapter 7). The innovative introduction of SSNNP can be carried out only at an international level and requires binding agreements shared by the majority of countries worldwide. At present, GNP is a major driver of environmental impacts and the decline of the ecosystem. United Nations could carry on (i) a parallel evaluation of the two indicators, with the purpose to make a comparative study of GNP and the SSNNP variations and identify gain and losses and, (ii) quantify the global "ecological debt accumulation" of current generations since the 1976 (Figure 8.3) and effects on future generations

- monitoring and reduction of most dangerous "externalities" (see Chapter 7), which represent an illegal system to produce huge profits, violating nature, life and people. The reduction of subsidies to fossil fuels and nuclear power is essential in making renewables progressively more competitive

- financing studies for a global sustainable development process taking into account that a standardized collection of statistical data on human progress, based on traditional and new indicators, should be adopted by International Organizations, the Banking System and Governments. This will not necessarily result in undisputable evaluations on the State of the World and the progress of Nations, but it will provide an adequate organization of reference data.

Humanity needs to know what should be done at the international and national levels to comply with a long term programme. United Nations Institutions should modernize their structure by reducing unnecessary bureaucracy, eliminating overlapping activities and initiating a new pathway. Our currently overpopulated World requires now the strengthening of democracy, as an internationally shared system based on common laws and rules. This process could help humanizing the economic system by convincing the supporters of capitalism that there is no more time to waste.

Deregulation is still a key principle of capitalism, as the only way to revitalize declining economies by attracting investors and maximizing profit. The most formidable obstacle remains the business-as-usual attitude which has been the rule for centuries, the justification for innumerable successful market-based initiatives, and the driver of the current global crisis. The need to make democracy coexist with capitalism remains a difficult but unavoidable task. Both components should undergo a full revision of fundamental principles for a common survival, identifying systemic inconsistencies and contradictions and sharing a new pathway for an acceptable coexistence.

According to Jay Haston[126] capitalism and democracy, however, are incompatible, as they are now, in their terminology, calculations and assumptions. As an example *"when economists claim the market is efficient, they actually mean the efficient distribution of benefits, not the efficient use of materials"*!

The Involvement of Nations in the Reference Transition Model

The RTM 2010–2040 includes three components respectively described in Chapters 10, 11 and 12:

- the *advanced energy revolution* (**E/R**), as proposed by EREC and Greenpeace in 2010, based on efficiency, the transition to renewables, the phasing out of the most polluting sources and the preservation of a natural capital that cannot be regenerated in decades

- the *sustainable development of infrastructures* (**SDI**), dimensioned at a regional level and implementable by nations on the basis of advanced guidelines for redesigning and planning the new society

- the *cultural revolution* (**C/R**), which is the indispensable support basis to tackle the current emergency, by enhancing the knowledge of people at all levels and fostering education and participation.

The implementation of the E/R based on renewables, efficiency and renewables, and the national planning for a SDI can be carried out by the existing institutions, which in a number of nations are already at work for a sustainable development process which addresses the basic concerns of environment, resources, poverty and democracy.

---

126 Jay Haston, *From Capitalism to Democracy,* www.dieof.org

The cultural revolution implies a great transformation of the social and economic domains, the change of lifestyles, the adaptation to a globalized world, the protection of environment and ecosystem, a safer life for present and future generations. Compared with the complexity of our current development and problems associated with a sustainable society, most people and leaders are more culturally inadequate, than their predecessors were before World War II.

The cultural revolution should prepare people in general to understand the current emergency, and — through a high level education — qualify political leaders and managers to cope with socio-economic problems. The process should involve the entire society from children to adults, taking care of the new global implications and the people's awareness needed to deal with the RTM and the SDP timing.

Since nations are not developed at the same level, programmes and projects should be considered first at the level of the USA, China, India, Russia and the European Union, which are supposed to hold adequate technologies and economic potential.

The three interactive components should be timely implemented and the related projects be defined on the basis of a holistic approach in which environment, energy, climate, agriculture, regional development[127] and advanced education play the major role.

The basic concept is that a planned "transition" is needed, not an abrupt change. Thus the current trend towards a more competitive free-market economy, based on moderate efficiency and growing profit, but indifferent to environmental and social issues, in the absence of a shared programme and global rules, is the way that leads to a global collapse.

---

127 Regional development should include: (i) modernization of the urban traffic systems through electricity-based vehicles, (ii) rehabilitation of the existing urban housing system to be energy efficient and based on renewables (wind, solar photovoltaic, geothermal and others), (iii) waste recycling, (iv) adoption of fully-integrated planning in new settlements, and, (v) cultural and science-based education for a new era.

CHAPTER 10

The Energy Revolution

Energy Sources and Carbon Dioxide Emissions

Energy is available today in huge quantities and types and is essential to the survival of our complex society. Figure 10.1 shows the potential of renewable energy sources of the World and the CO_2 variation 1960–2010 associated with the consumption of fossil fuels[128].

For the first time in the biological history, enormous quantities of energy are used for human activities, in turn drivers of progress and huge environmental impacts.

The primary energy derives (i) from finite non-renewable sources, as fossil fuels (coal, oil, gas) and radioactive minerals that cannot be regenerated at the current speed of consumption, and (ii) renewable energy sources (considered

---

128 In 2008, global energy consumption was evaluated at 15 TW (86.66% from fossil fuels, 6% from nuclear and 7.44% from renewables).

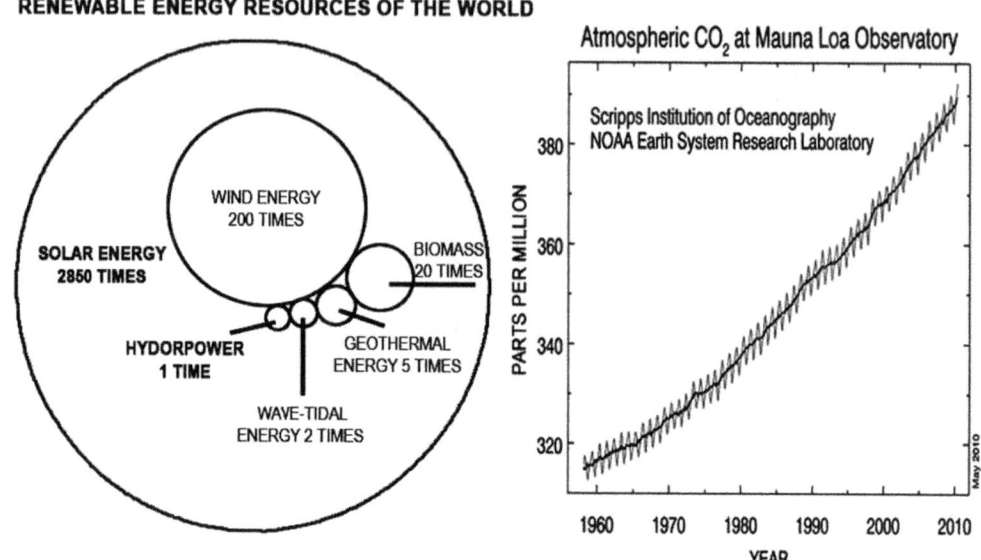

Figure 10.1 – (left) All renewable energy sources are 3,078 times greater than the current global energy needs per year. The technically accessible percentage, however, is much less, but still sufficient to support humanity. Adapted from WBGU, from the energy (r)evolution Report 2010; (right) CO_2 variation during the period 1960-2010 from Recent Mauna Loa CO_2.

as unlimited for practical purposes, since regenerated by the natural environment), which include wind, hydropower, tides, geothermal heat, ocean waves and wood.

The *theoretical potential* of all renewable energy sources is 3,078 times greater than global energy needs per year. Compared to the global primary energy demand (in 2007 of 503 EJ/a, IEA, 2009), the *total technical potential* of renewables at the upper limit would exceed, however, global demand by a factor of 32 only. The *theoretical potential of solar power*[129] alone is 2,850 times greater than World yearly needs, but the *technically accessible percentage* (TAP) is only 6 times greater than our society currently requires.

By taking into account all restrictions (legislative, technological, structural, ecological, economical, geographical), the TAP of solar power is sufficient to support

---

[129] The average solar energy that reaches the Earth is around $1KW/m^2$.

human needs for a carbon-free world, with advantages in terms availability, security and reduction in CO_2 emissions (which increased from 317 ppm in 1960 to 390 ppm in 2010).

According to Wikipedia[130], investment in renewables, based on global indicators, increased from 104 billion US$ in 2007 to 150 billion US$ in 2009.

Two factors alter the market of renewable sources: subsidies for fossil fuels and nuclear power and externalities[131] associated with their consumption; both these economic advantages make renewables to appear more expensive!

The 2007 Evaluation of Different Scenarios

A comparative study of renewable energy in global energy scenarios was carried out in 2007 by J. Hamrin, H. Hummel and R. Canapa under the Direction of the IEA and REDT[132] implementing agreement. The purpose was to explain the different shares of renewables in different scenarios and verify the assumptions critical to the use of renewables in future scenarios.

The evaluation involved:

- the goals of scenarios
- the role of renewables in terms of shares and growth rates
- methodologies used to build up scenarios and key assumptions
- cost, benefits and potential of renewable energy technologies
- main common bases and assumptions.

Humanity's energy future was considered to be the result of choice, not of fate; by providing different energy mix types and technologies, energy scenarios were taken as a basis to evaluate options and implications. Figure 10.2, compares the nine most

---

130   Source Renewable Energy (Wikipedia).
131   Externalities are costs not included in the full cost of the production process, but heavily harmful to the environment and ecosystem.
132   IEA: International Energy Agency, RETD: Renewable Energy Technologies Deployment.

documented scenarios, the role of renewables, the methodology used and costs and benefits.

REVIEW OF RENEWABLE ENERGY IN GLOBAL SCENARIOS (JUNE 2007)

	Type of Technology	IEA WEO 2006	IEA ETP	IPCC	WETO H2	WEC GES to 2050	EREC/ Greenpeace	EET Eff and RE	ASES	WBGU
Renewable Electricity Sources	Hydropower	X	X		X	X	X	X		X
	Biomass	X	X	X	X	X	X	X	X	X
	Geothermal	X	X		X		X	X	X	X
	Solar	X	X		X	X	X	X	X	X
	Solar Photovoltaics	X	X		X		X	X	X	X
	Concentrating Solar Power	X	X		X		X		X	X
	Wind Energy	X	X		X		X	X	X	X
	Ocean Energy	X	X				X	X		
	Other Renewables			X		X				
	Energy Efficiency	X	X	X	X	X	X	X	X	X
	Solar Heating and Cooling		X				X	X		X
	Biofuels	X	X		X		X	X	X	
	Nuclear	X	X	X	X	X		X		X
	Hydrogen	X	X		X			X	X	X
	Carbon Sequestration	X	X	X	X				X	X

Figure 10.2 – Review of renewable energy in global scenarios. Each scenario is based on a different energy mix. Seven out of nine scenarios follow the "business as usual" approach, while two — EREC/Greenpeace and WBGU — focus on a renewable energy mix for the reduction in CO_2 emissions. Source: Review of Global Energy Scenarios, by J. Hamrin, H. Hummel and R. Canapa under the Direction of the IEA and REDT.

The purpose of the report was to help policy makers, investors and consumers choose the best option. The 2007 evaluation concluded that:

- within scenarios based on the "business as usual approach" (seven out of nine) *the role of renewables in the future is expected to be insignificant, around 11-15% of primary energy production by 2050*, a mix which would still require the use of 89–85% fossil fuels in 2050

- two "non business-as-usual" scenarios, which assume new policies and technological innovations with the potential to change Humanity's future, are those proposed by EREC/Greenpeace[133] and the WBGU B1-400[134]. The two scenarios are considered as the *"most ambitious"* among the 9, as achieving a share of 50% renewables by 2050.

- electricity production from renewables reaches 70% by 2050 in the EREC & Greenpeace *"energy (r)evolution"* scenario, which is therefore considered as *"strikingly optimistic"*, compared to others

- only 4 of the 9 scenarios provided input cost assumptions for renewable energy technologies: those by the IEA-ETP[135], IEA World Energy Outlook 2006, EREC/Greenpeace, ASES[136].

Based on the above conclusions, the *"energy (r)evolution"* scenario has been considered by the Author the most adequate for humanity's future since including:

- policies and innovations, with the capacity to change present trends, new technologies and the use of renewables for electricity generation

- the centrality of climate change and the respect for the environment

- efficiency in the energy sector as a primary task

- carbon-free technologies, renewables and a production system compatible with the stability of the biosphere

- new laws to (i) phase out polluting sources and eliminate related subsidies, (ii) reduce externalities, (iii) establish global equity.

Among new scenarios published by EREC/Greenpeace after 2007, the *advanced energy (r)evolution scenario 2010* and the comparison with the 2009 *reference scenarios* by IEA WEO are described in this chapter.

---

133 The EREC/Greenpeace scenario was published in 2007. EREC is the European Renewable Energy Council, founded in 2000 by the European Renewable Energy Industry. Greenpeace is a well known non-governmental organization for the protection and conservation of the environment. Reports by EREC/GreenPeace were published in 2006, 2008 and 2010

134 WBGU, the German Scientific Advisory Council on Global Change.

135 IEA: International Energy Agency, ETP: Energy Technology Perspectives (2006), WEO: World Energy Outlook

136 ASES: American Solar Energy Society.

The *energy revolution* approach is a fundamental component of the RTM by concerning renewables, efficiency, global warming, climate change and new policies and procedures.

The Adv Energy (R)evolution Scenario 2010 by EREC/Greenpeace

Assumptions, key principles and new policies

The importance of energy requires no explanation, since energy is the source of everything in the Universe, the novelty today being the unprecedented and threatening use of energy by our species during the past two centuries. The *advanced energy (r)evolution scenario 2010* devotes particular attention to human development, which in the last six decades involved a greater consumption of non-renewable energy sources, and, to a lesser extent, of renewables, which have not increased at the same pace. Renewables are carbon-free sources, therefore their use is now appreciated as a unique opportunity to control global warming. Non-renewable energy sources include mainly fossil fuels which have been instrumental to the current stage of human civilization and the global warming.

In recent decades, the growing use of non-renewable sources carried negative side effects like pollution and other impacts; thus people, investors and consumers are seriously considering now different choices to support a carbon-free sustainable development.

Fossil fuels of course are still essential today, since the entire production sector is shaped on them, but growing environmental impacts are causing additional costs that people, municipalities and governments consider unbearable. Since 2007 — when the first study was published — energy (r)evolution scenarios by EREC and Greenpeace dealt with natural and human factors, on the basis of five "key principles":

- *respect for the natural limits of the environment*

- *phasing out polluting and unsustainable energy sources*

- *implementing renewable solutions, especially through decentralized energy systems*

- *creating greater equity in the use of resources*

- *decoupling economic growth from the consumption of fossil fuels.*

The scientific approach and technological innovations which qualify E/R 2010 scenarios are key cornerstones to contrast climate change and support sustainable development through the implementation of new social, political, administrative, and legislative policies.

Policy changes included in the E/R 2010 to stabilize climate are:

1. *phasing out all subsidies to fossil fuels and nuclear energy*
2. *internalizing the external (social and environmental) costs of energy production through 'cap and trade' emission trading*[137]
3. *mandating strict efficiency standards for all energy consuming appliances, buildings and vehicles.*
4. *establishing legally binding targets for renewable energy and combined heat and power generation.*
5. *reforming the electricity markets by guaranteeing priority access to the grid to renewable power generators.*
6. *providing defined and stable returns for investors, for example by means of feed-in tariff programmes.*
7. *implementing better labelling and disclosure mechanisms to provide more environmental product information*
8. *increasing research and development budgets for renewable energy and energy efficiency.*

According to the *adv energy (r)evolution* 2010 Report, conventional energy sources are subsidized on a yearly basis in the amount of 250–300 billion US$, nearly twice the financial support given to renewables.

This generates a market distortion, but also slows down the diffusion of renewables, which are artificially made more expensive.

The cost of externalities — ranging from more expensive medical assistance to the cost supported by municipalities for remedial measures against impacts — is central to the *energy (r)evolution*.

---

137 In the attempt to lower GHG emissions and reduce the impact of externalities the 1997 Kyoto Protocol called for 37 industrialized nations to decrease their emissions between 2008 and 2012 (5% lower than 1990 values), by fixing targets and timetables and by establishing a system of tradable pollution permits.

The advanced energy (r)evolution versus the reference scenario

In Figure 10.3, the EREC-Greenpeace *adv E(R) scenario 2010* and the *REF 2009* IEA-WEO global scenario for the period 2007–2050 are compared. The former is based on growing efficiency and renewables, the decline of fossil fuels and nuclear, and the decrease in CO_2 emissions by more than 80% by 2050. The latter, originally covering the period 1990–2035 but extrapolated to 2050, remains mainly based on fossil fuels and, as a consequence, on growing emissions of CO_2.

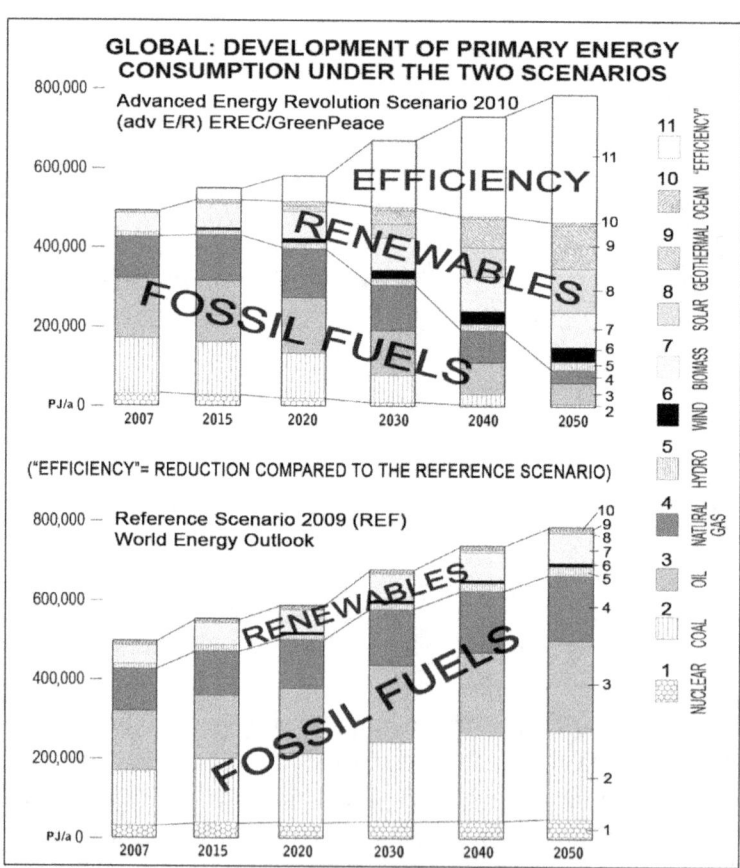

Figure 10.3 — The comparison between the adv E(R) and the REF global scenarios is self-explanatory. The rising consumption of fossil fuels in the REF would increase carbon dioxide emissions by more than 60% in 2050, thus becoming incompatible with the reduction of the mean temperature and the sustainable development process. Adapted from: energy (r)evolution, 2010, A sustainable World Energy Outlook, *by EREC and GREENPEACE*

At the global level, in 2050 the adv *E/R* is based on 41% efficiency, 47% renewables and 12% fossil fuels, compared to the REF scenario which foresees an increasing consumption of fossil fuels (coal, oil and natural gas) in 2050 amounting to 78%, plus 22% of nuclear energy and renewables.

The adv E/R scenario basically relies on the international action by UN and other organizations, a strong common involvement of public and he private sectors, and the participation of consumers. In this perspective, it goes far beyond the concept of revolutionizing the energy sector since the variety of new policies involved affect the society as a whole, the introduction of new technologies, the reshaping of production and consumption systems and the change of lifestyles.

The adv E/R 2010 scenario is strongly centred on the updating of free-market policies at the international level and the phasing out of externalities and subsidies to traditional energy sources.

Thanks to the variety of new technologies and combinations used in connection with renewables, the *adv E/R 2010* is expected to increase employment, the number of jobs achievable in 2030 being around 10.6 million, nearly two million above the REF scenario. An interesting "compendium" in the 2010 E/R Report for policymakers describes the major features of the advanced energy (r)evolution scenario through the period 2010–2050, based on "policy", "results" and the related support of authorities and institutions[138] responsible for implementing E/R policies and activities. The "cooperation framework" between institutions needs to be fostered and new international and national rules adopted and enforced. This implies a more adequate educational approach and advanced learning at all levels, if changes are to be in line with

- the emergency concerning climate change, degradation of the environment, food production, rising food prices etc.

- the transition to renewable energy sources which in turn carries the structural transformation of the society (urban areas, energy security and decentralization and transportation and industrial production) .

Global population will probably stabilize around 8.5–9 billion by 2050, thus the current and the next decade are to be regarded as an unprecedented emergency period. GDP (PPP) in 2010 reached $15.2 trillion in EU, $14.7 trillion in US and

---

138 The institutions involved are the UNFCCC (United Nations Framework Convention on Climate Change), the G8 and G20, national Governments, industry, consumers, regional governments and the car industry.

THE GLOBAL ENVIRONMENTAL CRISIS

10.08% in China, all together representing the 52.5% of $ 76.2 trillion of global GDP (Table 6.1). Under current conditions, natural and man related impacts, local wars and a variety of epidemic diseases are likely to develop uncontrollably in the near future, GDP, pollution and global warming reaching levels far higher than now, while the quality of life will inevitably deteriorate. Figure 10.4 shows the global trend in CO_2 emission reduction (from 27.4 million tonnes in 2007 to 4 million tonnes in 2050) under the adv E/R scenario 2010, as a result of savings from efficiency and other sectors (industry, transport and electricity). Emissions of CO_2 rise instead considerably by 2050 in the REF IEA/WEO scenario 2009. Data are not included, however, since the next paragraph is entirely dedicated to the OECD[139] IEA-World Energy Outlook 2010 report, which describes in detail the reduction in GHGs emissions within the 450-Scenario[140].

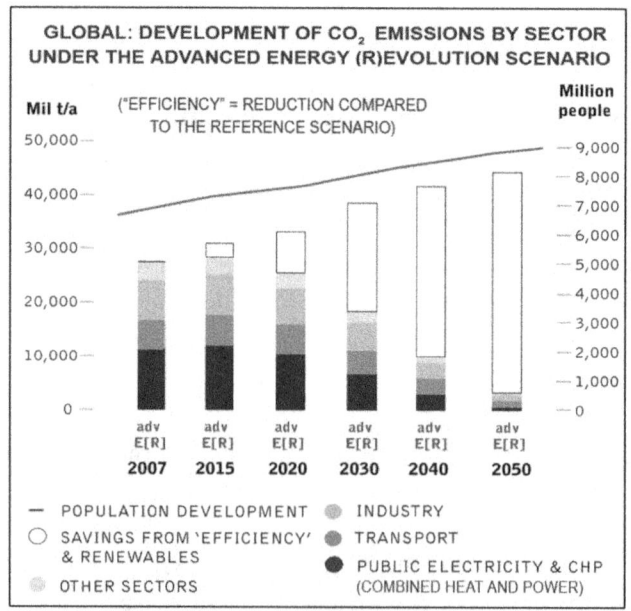

Figure 10.4 – Thanks to savings from efficiency, Global CO_2 emissions decrease from 28,400 Mil t/a in 2007 to 3,700 Mil t/a in 2050. Adapted from: energy (r)evolution, 2010. by EREC and GREENPEACE.

---

139 OECD: Organization for Economic Cooperation and Development, based in Paris; it includes 30 developed countries.

140 The 450 Scenario is based on the 2035 target of CO_2 concentrations remaining below the 450 ppp, which is considered the limit at which emissions should stabilize and halt global warming before it reaches 2°C, provided that New Energy Policies are adopted on time.

Figure 10.5 illustrates the regional OECD Europe scenario, again comparing the adv E/R 2010 the REF scenario IEA/WEO 2009.

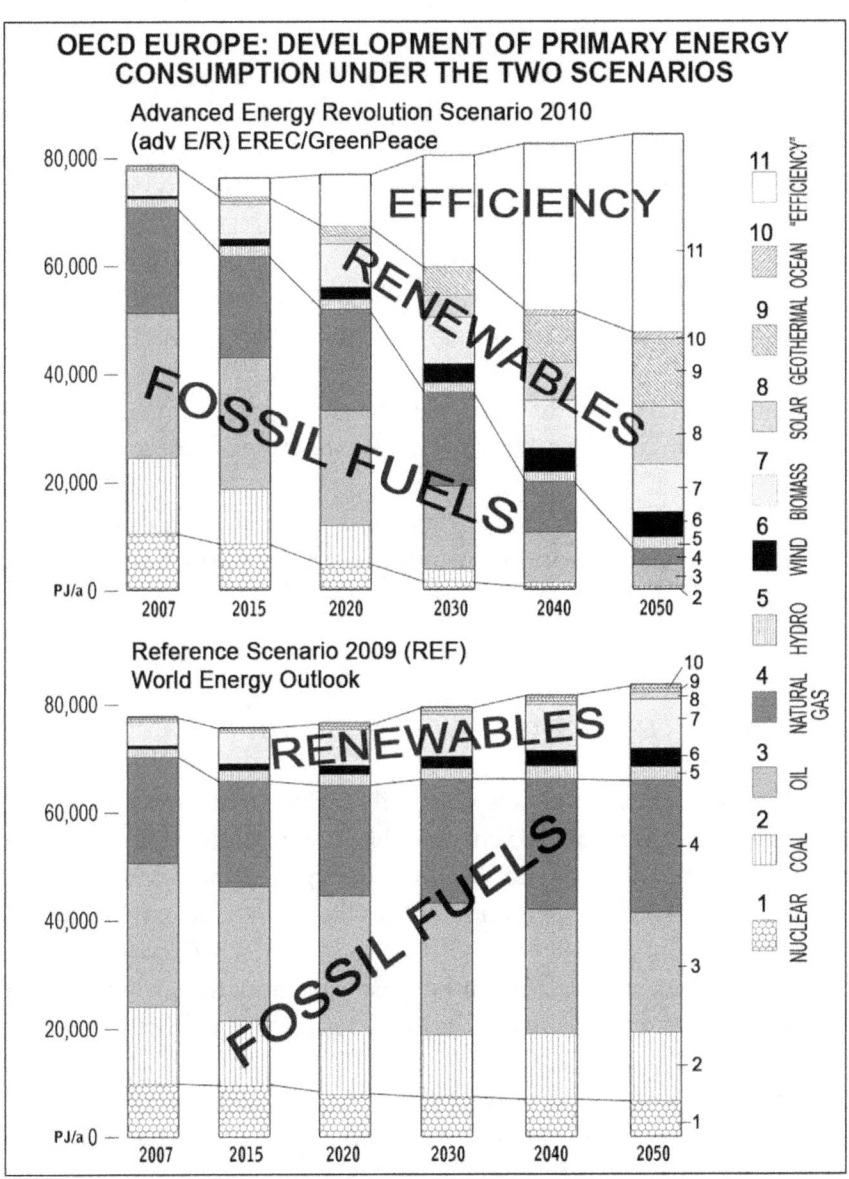

Figure 10.5 – Comparison between the adv E(R) and the REF in OECD Europe. Primary energy use decreases from 78,000 PJ/a in 2007 to 52,000 PJ/a in 2050 in the adv E(R) scenario, while remaining stable in the REF. Adapted from: energy (r)evolution, 2010. A sustainable World Energy Outlook, by EREC and GREENPEACE.

The adv E/R scenario develops efficiency and renewables which account for nearly 92%, with fossil fuels at 8% of the total energy consumption in 2050. By comparison, the mix of the REF scenario changes slightly in time remaining around 80,000 PJ/a, out of which renewables reach 23% and fossil fuels plus nuclear 77% in 2050. Major aspects of the adv E/R Report 2010 include the target of zero deforestation at the global level in 2020, the efficiency target of 20% renewables in 2020, and the phasing out of coal and nuclear sources within 2030.

The report devotes attention to the costs and benefits analysis providing essential indications on:

- energy types and mix, the costs of new technologies for different solutions, cost and benefit estimates

- regional scenarios on the basis of 10 world regions, for which global energy statistics were developed by the IEA (International Energy Agency)

- pros and cons of major energy types and a database of global (and regional) electric energy generation, installed capacity, energy demand and CO_2 emissions targets.

Some final considerations on the adv E/R scenario 2010

The *adv E/R scenario 2010* represents not only a way out of the emergency, but also an opportunity for unprecedented changes. Unless humanity finds ways of protecting the environment and ecosystems and controlling population growth, human activities and impacts, the emergency will make the global environmental crisis to reach earlier its tipping point. According to the 2004 scenario by D. Meadows et al., the decline of resources, food and industrial production, followed by the collapse of population and the rise of pollution (Figure 9.1), are likely to reach critical points between 2020 and 2035.

In case the emergency were to worsen, further social progress would be halted, and our chances of surviving decorously would begin to decrease under the combined effects of climate change, the collapse of the free-market and the reduction in the services provided by the ecosystem. Efficiency could turn out to be the major source of energy (Figure 10.3), provided that adequate standards are implemented in regulating consumption and fixing the limits of household appliances that heavily contribute to making energy bills rise.

Binding targets for renewables are fundamental to establish common rules, renew the global production system and lead to higher efficiency. The transition to an economy based on renewables and new technologies implies indispensable changes among which connections to new grids.

To foster the rise of renewables, stable returns need to be granted to investors. Some specific problems within the adv E/R 2010 need to be mentioned.

First, in January 2010 there were 436 worldwide operating nuclear power plants, that is eight less than in 2002. During the last 10 years 32 new nuclear plants have started operating, the majority of which in China and India. Should nations massively turn worldwide to nuclear, it would mean making huge investments in the sector and freezing the progress of renewables and the potential for new clean technologies for the next two decades or so.

The phasing out of nuclear (foreseen by the *adv E/R 2010* within 2030) will prove a difficult task. Nations like the USA, France, Japan and others (among which some emerging powers that have already invested in nuclear energy), will most certainly encounter difficulties in effecting the change. Moreover new investments in nuclear energy would end up by preserving a costly inheritance for future generations.

Fukushima's nuclear disaster (March 11, 2011), triggered by a very strong earthquake and the associated tsunami, partly froze new programmes in the sector, heating the debate on safety measures and the updating of several old-generation nuclear plants. On top of these problems, worth mentioning is the Yucca Mountain Nuclear Waste Repository (Nevada) which, due to the opposition by local authorities, was terminated in 2010 after more than one decade of investments. Nations not subject to seismic risk are likely to keep nuclear beyond 2030. At the end of May 2011, the Swiss and German Governments announced the phasing out of their nuclear plants.

Second, although the reduction in CO_2 emissions was shared by the majority of countries, the phasing out of coal by 2030 will not be easily achievable. The current energy development trend of China and India and other countries is still based on coal. The reasons are: (i) coal-fired plants are much cheaper if compared to other technologies, (ii) coal technology is based on a long term experience and, (iii) the demand of electricity is rising fast. Around 2020, the consumption of coal could be on a decreasing trend in some developed countries and growing in emerging nations. Coal is available in huge quantities, so in the case of energy shortages it would turn out to be the cheapest non-renewable source.

Third, the growing use of biomass is expected to stabilize at 12% in the period 2030–2050 (adv E/R global scenario, Figure 10.3), the potential remaining high in

Latin America, Europe and in some transitional economies. The adv E/R 2010 definitely implies a complex implementation process, and an intergenerational perspective in which present and future generations should be able to share equal opportunities. E/R is not limited to renewables and efficiency targets, since it involves also energy sawing and, above all, the transition of our society towards clean energy sources, decentralized energy systems and co-generation technologies[141] which today are already available in the market.

A further reason for adopting the adv E/R 2010 is the foreseen increase in employment levels, which are currently declining everywhere except in emerging countries. Rich nations need to introduce structural changes in their energy, economy and production systems to recover market competition and create new jobs.

The World Energy Outlook 2010

Tackling climate change and securing a sustainable future

The WEO–2010 Report by IEA has been conceived in a condition of global uncertainties, among which (i) the risk of a growing climate change, (ii) the difficulty of stabilizing the energy sector under the threat of energy shortages and rising prices and, (iii) the financial global crisis exploded in 2007. The report highlights the importance of energy as the bearing structure of the economy worldwide, recommends a rapid transition to renewables and demands the reduction of GHGs emissions, which are considered drivers of climate deterioration and humanity's increasing difficulties. Basic key-issues of the WEO 2010 Report are:

- major challenges are today global warming, climate change and a safe and sustainable energy system to be implemented through a rapid transition to a low-carbon society

- the vital need to limit the mean temperature rise to 2°C and a maximum level of 450 ppm of CO_2

- identification and implementation of government commitments before 2020

---

141 Cogeneration technologies involve heat engines or power stations designed to produce electricity and heat at the same time.

- innovative policies and technologies

- the removal of subsidies and incentives, and effects on government budgets and climate

- the unprecedented role that renewables and efficiency can play during the transition and the related cost

- the oil and coal consumption forecast in 2035 and the key role that China and India will play in the global energy market

- the consequences of failure in the case that new policies and technologies are not implemented on time.

The main problem for a fast transition to renewables is the reluctance and delay of governments in pursuing sustainable development.

The puzzle of components and the IEA-WEO 2010 major scenarios

Before starting to evaluate scenarios, some specific difficulties need to be mentioned:

- oil and other fossil fuels are the major cause of direct and indirect environmental deterioration and harmful impacts on life and the environment. The major threat posed by the transition to renewables is that the World's economy and production are structured on the basis of fossil fuels. Transforming the system, which was mainly developed during the last 60 years of cheap oil prices, at the speed imposed by climate change, represents an enormous challenge and implies huge investments.

- the complexity of the transition is determined by the fact that fossil fuels remain a dominant source, although quantitatively decreasing until 2035, while energy from renewables should increase in order to compensate the system and make it work for a rising population of around 8 billion in 2025.

The overall picture which emerges from the WEO scenarios is that population, economic growth and impacts act as almost uncontrollable drivers, in conflict with the preservation of the environment and the ecosystem, and ultimately the human progress.

Our species has suddenly realized the need for a change; this implies the transition from an instinctual behaviour to a reason and technology-based action that can ferry humanity out of the crisis. Contrasts among nations and political parties are unavoidable, and humanity is being involved in a sort of "mission impossible" that should be, however, timely accomplished!

The World Energy Outlook 2010 recognizes the challenges and timing associated with climate change, the effort made by the UN International Conferences to implement policy commitments, and the new role of the emerging economies in shaping the energy sector.

Two major scenarios are proposed by the World Energy Outlook 2010 in Figure 10.6.

The *new policies scenario* which

- assumes a slow implementation of the commitments announced under the Copenhagen accord and plans, and the extreme difficulty of keeping the global temperature below 2°C,

- provides benchmarks to control achievements and limits in relation to climate change,

- foresees that under current trends the concentration of GHGs will rise above 650 ppm CO_2-eq and temperatures will reach over 3.5°C in the long term and,

- foresees that if strict and more ambitious targets are fixed, a faster removal of subsidies is agreed by the G-20 and efficient policies are implemented within 2020 compared to those agreed in Copenhagen, the target of 2°C as an upper limit can be maintained.

The *450 scenario* which is based on

- a strict implementation of new policies and binding reductions in GHGs emissions; the scenario is practically centred on a shared abatement of gaseous emissions

- more expensive technologies involving efficiency, renewables, biofuels, nuclear and CCS, with the potential to limit GHGs concentrations to about 450 ppm (CO_2-eq) and keep the average temperature below 2°C.

THE ENERGY REVOLUTION

These benchmarks are considered by the IPCC the limit beyond which temperature could rise uncontrollably. The CO_2 reduction of 20.8 Gt by 2035 (7.1 Gt reduction under the New Policies Scenario plus the cumulative 13.7 Gt reduction under the 450-Scenario) are believed to stabilize climate.

The 450-scenario is in practice a more advanced and expensive version of the New Policies Scenario, with the fixed time target of 2020 for a drastic but progressive slowdown of greenhouse gasses emission.

The concept underlying the 450-scenario is centred (i) in the belief that humanity is facing an ultimate chance, as pressed by the demand of growing electricity production more easily achievable through conventional coal-fired plants, and (ii) the assumption that the target of not trespassing 450 ppm CO_2 equivalent can be achieved by 2035, by adopting and implementing new more expensive technologies.

Figure 10.6 shows World Energy Outlook 2010 scenarios.

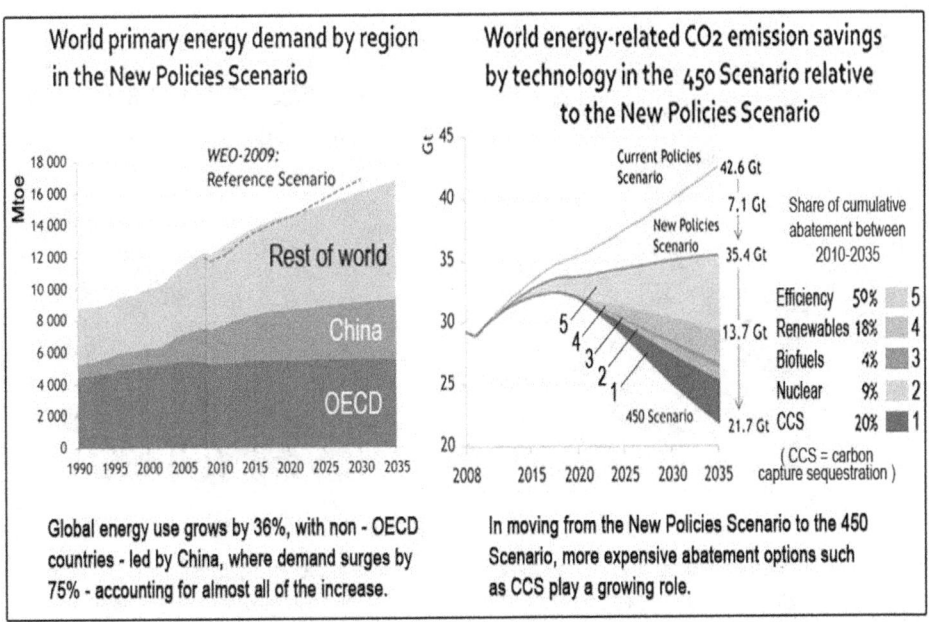

Figure 10.6 – (left) World Primary energy demand by region in the New Policies Scenario; (right) the 450-scenario based on World energy-related CO_2-eq emission abatement by technology. Source: World Energy Outlook 2010, Presentation made by Mr. Tanaka in Beijing on 17th November 2010, © OECD/International Energy Agency, adapted for presentation purposes.

The oil peak and coal-fired electricity generation

Figure 10.7 (left) shows that the "oil peak" of currently producing fields was reached between 2005 and 2008. However, by hypothesizing "oil fields yet to be developed" and "oil fields yet to be found", the New Policies Scenario envisages the possibility that the 2005–2008 "crude oil peak" could remain constant at 68 mb/d until 2035.

In the meantime, natural gas liquids and unconventional oil are expected to increase until 2035, when a global oil production of 96 mb/d will be reached. This would imply a considerable rise in CO_2 emissions and a further rise in the mean global temperature.

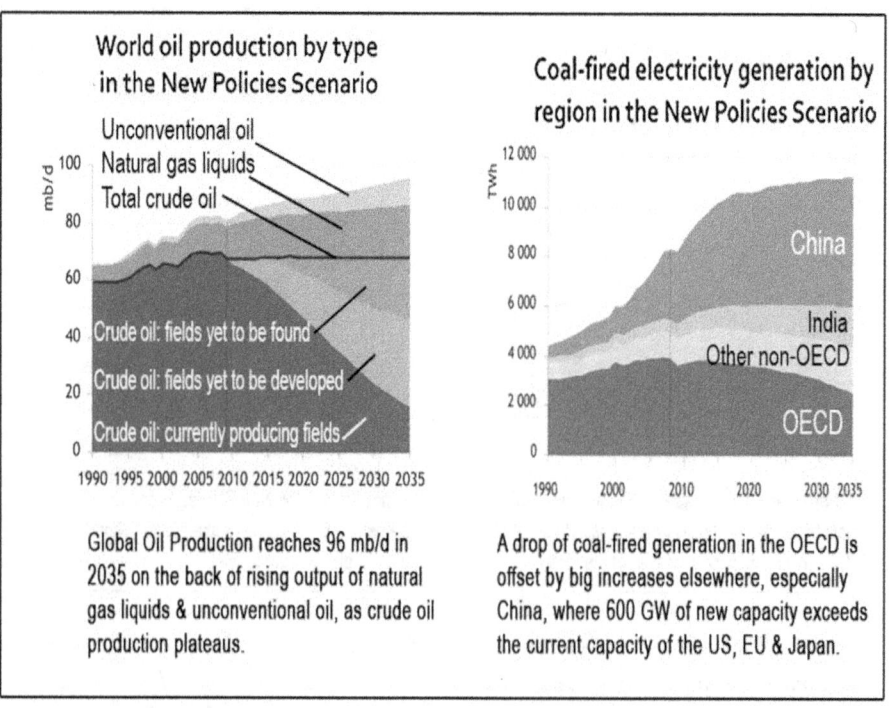

Figure 10.7 – World oil consumption and electricity-related coal use in the New Policies Scenario 1990-2035. Source: World Energy Outlook 2010, Presentation made by Mr. Tanaka in Beijing on 17th November 2010, © OECD/International Energy Agency, adapted for presentation purposes.

In sum the IEA-WEO 2010 Report overcomes the negative impact of an oil consumption amounting to 96 mb/d in 2035, by proposing the more ambitious *450 Scenario,* based on the drastic abatement of emissions. The view of global oil production rising to 96 mb/d in 2035 (Figure 10.7 left) was widely criticized, as not supported by data concerning "oil fields to be developed and oil fields to be found". Nevertheless, some significant quantities of oil can still be available in unexplored land areas and the ocean floor.

The 2008 study *"Crude Oil – The Supply outlook"*[142], by J. Schindler *et al.*, showed that in former OPEC and FSU (Former Soviet Union) the peak in oil producing countries was reached in 2006, with around 38 Mb per day. The study demonstrated that the 17 biggest oil companies in the World in 1997–2007 were unable to increase their aggregate oil production (as the drop[143] in *crude oil from currently producing fields* since 2008 shows), coincidentally in a period of unprecedented growth of prices.

The perspective of a rising global warming, however, and the associated faster melting of the Arctic ice cap opened the race among Denmark, Sweden, Iceland, Norway, Russia, America, and Canada to extend their territories beyond current boundaries. Extensive ocean bed mapping is going on now, and probably ¼ of the still undiscovered deposits of oil and gas could be hidden in the Arctic Ocean floor[144] and/or elsewhere.

According to recent news, Russia may become the biggest gas and oil producer during this decade.

Thus, the real "global oil peak" could be probably reached around 2020–2030 or so, if additional oil is discovered. Should this happen, humanity would remain trapped in the nightmare of oil still available for years, which would make GHGs emissions grow further and the transition to renewables proceed more slowly.

Natural gas liquids and unconventional oil from Canadian Athabaska's sands will boost the global oil production, but prices will inevitably rise until 2035, unless new policies are rapidly and strictly implemented.

In my opinion, the end of "cheap oil" is near, huge investments and time are needed before the oil from the Arctic Ocean floor and other areas is discovered and commercialized. Moreover, after the BP oil spill in the Gulf of Mexico in 2010, oil

---

142   Source Oil Report: www.energywatchgroup.org

143   The unproved existence of crude oil yet to be developed and yet to be found means (i) that oil companies and oil producing countries do not like their classified information to be divulged, preferably keeping the secret over oil and gas reserves which can be commercialized at growing prices in the near future and, (ii) oil exploration is now including the continental platform and large unexplored areas worldwide.

144   National Geographic, May 2009.

from the deep ocean floor is more risky and expensive, as it is the extraction of heavy oil from the Athabaska sands in Canada.

The good news is that the increasing oil price should make renewables more competitive, while the bad news is that emerging and developing countries will increase the consumption of coal. In sum, oil and gas prices will rise[145], and coal consumption will grow faster than ever as a consequence of the demand of electricity.

Figure 10.7 (right) shows that coal consumption for electricity generation under the New Policies Scenario decreases in OECD countries from 2012 onwards, but significantly grows in China, India and other non-OECD Countries.

According to the article published by David G. Victor (January 2011 issue of Scientific American), China, India and Brazil are making a significant effort to control gaseous emissions and lower the pollution of most industrialized cities. China despite investments in nuclear power plants and renewables, however, is still using huge quantities of coal to preserve current economic efficiency and support the growing demand in electricity.

As one of the largest producers of coal the country is involved in a big construction programme of coal-fired plants, which will inevitably increase the emission of carbon dioxide into the atmosphere. China's GHGs emissions have already exceeded those of the USA in 2008 and the coal consumption of electricity is expected to grow at an accelerated rate between 2010 and 2020 as it did in the period 2000–2010. Should the foreseen demand for electricity be maintained until 2035, as shown in Figure 10.7, the consumption of coal in China would nearly double between 2010–2035, while in OECD countries it would decrease by 1/3 compared to 2009 values.

The growing use of coal in China, India and other nations is likely to thwart the global effort in emission reduction. China, India and other non-OECD nations — in the absence of an international effort to reduce GHGs emissions — will increase coal use for electricity generation by nearly six times by 2035 compared to 1990, China alone in the same period raising it eight times[146]. A slowly decreasing or stable availability of oil around 100 US$/barrel until 2020 or so, compensated by the growing use of coal will inevitably increase CO_2 emissions and delay the transition to renewables. In sum

- *commitments to the reduction of GHGs emissions* through energy saving and efficiency should be taken seriously into consideration at an international

---

145 In April 21, 2011 the oil barrel was sold at 106 US$.
146 China is relatively rich in coal, thus at least three new coal-fired plants are built every month, in a trend that will continue for at least a decade.

level, to prevent that part of the World accepts drastic reductions, while other nations continue to use rising quantities of coal

- *subsidies* distort the market's competition, do not improve energy security, encourage wasteful use, fuel adulteration and smuggling, and slow down the progress of renewables. In 2009 subsidies to fossil fuels reached 312 billion US$ (compared to 558 billion in 2008)

- *fossil fuels* will play the role of essential energy sources until about 2035, the associated environmental impact remaining huge, but still avoidable if new policies are implemented

- *emerging economies* (mainly China and India) will increasingly shape the energy domain and the World's economy.

Concluding Remarks

The *adv energy (r)evolution scenario (E/R)* 2010 remains deeply rooted in the free-market domain and based on the following concepts:

i. the approach is pro-active since it anticipates future problems and suggests solutions to be implemented now
ii. the "business as usual approach" in the energy domain is fostered by a permissive legislation, incentives and subsides to coal and oil and a total disregard for externalities; thus fossil fuels are artificially made more competitive than renewables
iii. the current climate trend has been altered through a variety of impacts resulting from human activities disrespectful of the biosphere, ecosystems and human rights
iv. guidelines for change require a giant effort in "energy efficiency", the transition to renewables and the phase out of nuclear and coal by 2030 and a progressive reduction of oil and gas consumption within 2050
v. a higher level interaction between the public and private sectors is indispensable in restructuring human society in the direction of sustainability.
vi. in the absence of timely actions, the World's economy will be reshaped by emerging powers, which in turn might neither be able to control GHGs emissions, nor curb climate change.

The adoption of the RTM implies a general trend towards a sustainable society, through a growing use of renewables, the increase of energy efficiency and the phase out of carbon-rich sources. An integrated planning for energy and infrastructures and a cultural revolution are essential in fostering people's participation in shaping the future. RTM in sum is an unprecedented proposal for an unprecedented global crisis.

The *WEO Report 2010*, with the 450-Scenario goes definitely beyond the traditional "business as usual", by recognizing the dimension of the emergency and the need for international commitments, if the abatement of GHG emissions is to be achieved. But uncertainties in the implementation of new policies are evident and the difficulty of reshaping nations with different development levels remains a giant obstacle.

Two unavoidable trends bar the sustainability process.

The *first major uncertainty* concerns the slower trend of humanity in gaining awareness of the crisis and in implementing new policies, compared to the fast pace of global warming and the associated effects. Too many nations are unable to cope with the complexity of the crisis and take decisions quickly.

This is often referred to as the lack of "political will", an expression which hides the difficulty of people who are mostly used to think locally while major problems grow globally.

The *second uncertainty*, undeclared but intuitively logical, is that a unilateral action by the big nations is urgently needed to unlock the current trend of letting things work out by themselves. According *The New York Times* (12/12/2011) the Durban Conference on Climate Change ended with the conclusions to shift to 2020 the binding agreement on GHGs reduction!

As a consequence of the Fukushima nuclear disaster in Japan (March 11, 2011), the debate on nuclear energy heated up worldwide[147]. Pro-nuclear activist point out that the cost of a kilowatt hour of electricity from nuclear (including construction, production and decommissioning) is five times cheaper than solar and that the space required by solar panels to produce the same energy of a nuclear plant is tens of times greater.

This broad comparison does not take into account (i) the risk of nuclear compared to other sources, (ii) the unsolved problem of nuclear waste, (iii) the cost of radioactive material which will almost certainly rise as oil did, (iv) the long term

---

147   By the end of May 2011 Swiss and German Governments decided to phase out nuclear by 2022. As a consequence of the June 12 , 2001 referendum, Italy banned for the second time nuclear.

maintenance of nuclear plants and the vulnerability of local deposits of radioactive waste.

Unless exceptionally new technologies (at the moment unavailable) solve once for all uncertainties in the nuclear energy domain, other nations will join the anti-nuclear movement. Today's contribution of nuclear energy to the World Energy consumption is 6% only.

The book *"Only One Earth – The Care and Maintenance of a Small Planet"* (1972) by B. Ward and R. Dubos, ends with a sentence expressing the bottleneck in which the Authors felt humanity was trapped at the beginning of 1970s: the millennial habit of taking individual decisions and the explosion of nationalism in open contrast with the unity of the biosphere and the interdependence of humanity and nature.

CHAPTER 11

The Sustainable Development of Infrastructures

The Pathway Towards a Sustainable Society

Sustainability in the ecological context is a process concerning environment and human systems and their ability to preserve a stable balance within the given conditions. The 1972 Rio de Janeiro International Conference on Environment and Development was the historical event in which global consensus and political commitment were for the first time considered indispensable for their integrative nature and the target: put at the centre of the debate humanity's future and the transition to sustainability. The task forty years later revealed to be a most formidable one, due to the innumerable diversities among nations and, above all, the lack of a shared knowledge about environment, sustainability and the action to be implemented. Within the RTM the sustainable development of infrastructures (SDI) is essential in planning activities, is complementary to the energy (r) evolution, is fundamental in involving Government Institutions, investors and local Authorities. Major sectors are:

- energy efficiency: buildings need to be based on photovoltaic systems, insulation and ventilation systems, solar thermal collectors, efficient thermal power stations

- urban development: comprehensive planning and management, adequate construction techniques that reduce pollution, land use and the adequacy of new and old structures to a decentralized energy future

- transport systems: high level design is required to minimize environmental impacts, take into account that current combustion engine cars will coexist for years with grid-connected electric vehicles

Above all, experts involved in integrated design need to share key principles of sustainability, the respect of nature and limits and a wide knowledge of new technologies for the development of a carbon-free efficient society. The idea that humanity can follow the current path, worsening the ecological balance of the biosphere, is short-sighted but real. Government planning policies, sustainable architecture, advanced knowledge in the design of infrastructures and cooperation among experts are key ingredients for a sustainable society.

Infrastructures for a Sustainable Development

To achieve sustainability and efficiency large to small-scale integrated infrastructures are needed, the first step being the timely implementation of national programmes involving public and private sectors, investors and end users. The design approach should combine different but coordinated solutions suitable for old structures which need renovation to efficiently coexist with modern structures. Innumerable papers, books and national and international conferences are dedicated to the integration of renewables, new technologies and advanced planning. Options for different needs and conditions are described in the adv E/R 2010, which provides suggestions for the transition to renewables and an increasing efficiency in urban areas in the perspective of a decentralized energy future. A sustainable programme for infrastructures[148] based on an environment-compatible social and economic development should include:

1. *the transition to renewable energies*
2. *efficiency*, which concerns traffic and vehicles, new designs of house-hold appliances, insulation of new and old buildings
3. *construction, renovation and updating of deteriorated infrastructures* based on (i) integrated planning for the development of renewable carbon-free energies from wind, solar photovoltaic, geothermal and hydroelectric plants, and (ii) developing pilot projects in urban and rural areas
4. *land use planning*, which pursues an ordered and regulated use of the territory, based on a comprehensive design for the achievement of community goals leading to securing environmental, economic and social stability
5. *transport and traffic.* The development of public transport electricity-based systems is the way to reduce the number of combustion engines worldwide, increase the number of non polluting-vehicles, and lower the current congestion in urban areas.
6. *improvement, maintenance and efficiency of water supply and sanitation systems.*
7. *protection of floodplains and river training activities.* Water management is essential in reducing the impact of floods at the same time enhancing the recharge of aquifers through small dams and reforestation.
8. *waste recycling includes* advanced technologies which allow a nearly full utilization of waste by recycling and the production of biogas.
9. *reforestation plans* are the natural way of capturing and storing carbon dioxide, at the moment (together with the efficiency in energy use and the transition to renewables) one of the most effective ways to reduce carbon dioxide concentrations in the atmosphere.
10. *a database of sustainability indicators* needs to be available worldwide, plus the monitoring of activities and the design principles of new advanced infrastructures in different local conditions.

As a consequence, the current organization of the public sector, responsible for the implementation of the SDI, needs to be thoroughly revised to interact with the private sector for the identification of national programmes and new technologically

---

148 An interesting paper is the "Policy Guide on Planning for Sustainability", 2000 (by the NGO APA, American Planning Association) which synthesizes information on dimensions of sustainability, indicators of unsustainability, steps in the planning process, policy positions and planning actions.

advanced solutions. The infrastructure sector has historically been instrumental worldwide in shaping urban areas, in conceiving huge projects in the domain of water supply and sanitation, in the construction of transport systems and power plants.

With the SDI a new pathway needs to be started by adopting a fully integrated multidisciplinary planning for an advanced design and construction of infrastructures which need to be physically, socially and culturally shaped for a sustainable society in which present and future generations share equal opportunities.

The past experience of a random-style design and construction of infrastructures and the absence of land use planning had devastating effects in Italy[149] and other countries; projects were often useless, inefficient and in a number of cases were not even completed. Overpopulated areas need particular attention to avoid unnecessary infrastructures which produce the deterioration of the territory and the degradation of good quality soils that take decades before a natural regeneration is achieved.

---

149 A huge flooding ravaged Tuscany and Liguria October 26, 2011, with mudslides, landslides and a huge quantity of debris, the disruption of the transportation system and casualties. The absence of prevention and environmental protection measures was fatal.

CHAPTER 12

The Cultural Revolution

Wisdom, Ignorance, Scepticism and Misinformation

In his 2008 paper[150] *"Target Atmospheric CO_2: Where should Humanity aim?"* J. Hansen concludes as follows: *"The most difficult task, to phase out over the next 20–25 years the use of coal that does not capture CO_2 is herculean, yet feasible when compared to the efforts that went into World War II. The stakes, for all life on the planet, surpass those of any previous crisis. The greatest danger is continued ignorance and denial, which could make tragic consequences unavoidable"*.

Hansen's worries were based on undisputable data: the level of 385 ppm CO_2 in 2007 was already too high, and the reduction to 350 ppm[151] very difficult to achieve by increasing the price of CO_2 emissions and at the same time phasing out coal, except in the case of CO_2 capture and storage (CCS).

Carbon tax and emission trading failed to generate the expected results, while CCS is still expensive. Today's CO_2 concentration (for which rich and emerging

---

150 Target Atmospheric CO2: Where Should Humanity Aim?
151 CO_2 350 ppm was the 1990 value.

countries share the major responsibility) is far above the maximum level of 287 ppm which was reached over the last one million years.

A further population growth during the next three decades means the need for a consistently greater food production, a change in dietary habits (in rich and emerging nations), a drastic reduction of waste, a strict control of food prices and an equitable distribution.

Growing population and poverty, and the global scarcity of food, will necessarily increase people's migration and cause uncontrollable conflicts, unless substantial changes are timely implemented. In no way humanity can provide cheap food for all, the growing price being not affordable by the poor.

Paradoxically it is easier to save energy, reduce waste and make a wiser and efficient use of primary resources, than providing a minimum decent food support to the growingly undernourished one billion people. Food should not be used irresponsibly to maximize profit! An example of lack of knowledge and awareness in the past three decades was the blind opposition to the UN birth control campaign.

By contrast, an excessive increase in wildlife compared with the stable (or decreasing) biocapacity of the environment is fully recognized[152].

The limitedness of environmental resources essential to life was ignored by the July 1968 Encyclical *Humanae Vitae*. In the 1980s some bishops stated that the Earth could feed 40 billion people, totally unaware that arable land and water for irrigation are limited and at that time already affected by problems. This information is provided by Paul Ehrlich in his article "Can we respond to the growing environmental threat to civilization" included in the book *"Environment in Peril"*[153].

Sceptics do not even recognize the connection between global warming, overpopulation and man-made activities, claiming that temperature variations occurred innumerable times in the Earth's history.

---

152 Vegetation and water are essential for herbivorous. In the September 2008 issue of National Geographic, Karen E. Lange described the death of 500 elephants caused by thirst and starvation during 2005 in Zimbabwe's Hwange National Park. Every day elephants eat 400 pounds of food (180 kg) transforming the ecosystem from wooded savannah to scrub grassland. When the number of elephants exceeds the biocapacity of the area, "managers may have to resort to shooting elephants to save ecosystems". This halts their reproduction, avoids poaching, and preserves the habitat for other herbivores. Our overpopulated World, with 7 billion people in 2011, will inevitably reach the limits of the system and – in the absence of a birth-control campaign and a consistent reduction of CO_2 emissions – the declining biocapacity is likely to turn into a global environmental catastrophe.

153 Environment in Peril, (1991), edited by A. B. Wolbarst, Smithsonian Institution Press.

What makes the current condition an unprecedented event is the role of humanity as the major driving force, the speed of environmental change, the decline of ecosystems and the potential occurrence of uncontrollable global disasters.

A radical denial of the global crisis was expressed by B. Lomborg in his 2001 book titled *"The Skeptical Environmentalist"*. Lomborg declared that the Earth was not in peril, the environment not undergoing degradation, resources were not depleted at a faster rate and forests worldwide were not disappearing. After his 2007 change of mind about the state of the environment, Lomborg stated that climate change is a problem, "but not the end of the World".

The Background of the Global Environmental and Financial Crises

The complex history of the last 100 years

The need for a cultural revolution (see Chapter 9) arises from the human difficulty in understanding the global environmental crisis, in defining causes, dimensions, short-term effects and new policies to prevent a catastrophic conclusion. Human inadequacy in sum is basically cultural in nature and is grounded on the unprecedented and fast-growing conflicts between human development and biosphere limits and dynamics. A short summary of the historical sequence of events during the last 100 years can help in understanding the origin and the complexity of current environmental crisis and the human limit in reaching a global consensus on urgent remedial measures.

The *first period*, from 1915 to 1946, was marked by two World Wars, the second of which ended with the victory of the USA, URSS, UK and France, soon after followed by the rise of the USA and URSS as global superpowers, the split of Europe and the rest of the World, either under the influence of America or the Soviet Union.

The *second period*, from 1946 to 1991, was characterized by the Cold War and several local conflicts which for 40 years saw USA and URSS engaged in an indirect confrontation in Asia, Africa, Central and South America. The fall of the Berlin Wall on November 9, 1989, soon to be followed, between January 1990 and December 1991, by the disintegration of the URSS, marked the end of the cold war. For 45 years, the US tried to halt the spread of communisms and the URSS attempted to control nations in which the influence of the USA was declining.

Concerning the economy, since 1950s, the global rise of the GDP was highlighted as the measure of the economic growth of Nations. The associated global environmental impacts and the widespread pollution of sea, air and land, which progressively increased through the years, were ignored. During 1970s the transition from a sustainable to an unsustainable World, unprecedented in human history, took place. Governments were neither aware of this big change, nor of combined effects of an ever growing economy in a rising unsustainable world. With the development of supercapitalism in 1980s internal contradictions inside the advanced western countries widened, almost inadvertently undermining the legitimacy of democratic institutions.

The *third* period, between 1992 and 2011, can be considered as the age of awareness, marked by the discovery of a fast rising global warming, the further deterioration of the environment and the spreading of globalization. The complex interplay of factors — marked by the explosion of emerging economies, the feared decline of rich countries and the unscrupulous behaviour of the banking system and investors — has triggered the global recession, which is in 2011 still heavily affecting advanced economies and in different ways the rest of the World.

Humanity seems now to be ready for a big social and cultural change, at the same time being forced to do it under unprecedented constraints.

The difficulty of changing the present and figuring out the future

The knowledge of events which unfolded during the past two decades, is the precondition to understand why the awareness of current environmental crisis is so recent, and the critical revise of capitalism and democracy so urgently needed. The relationship between globalization and democracy is the topmost issue at the beginning of the Third Millennium, by concerning the future of humanity and the fear of Western democracies on the one side involved in a difficult economic competition with the emerging countries, on the other worried for the advancing global environmental crisis. In sum never before survival involved an intelligent species, at the same time (i) aware of the environmental change, political, economic, social and cultural issues, in turn affecting democracy, governance, interdependence, and, (ii) worried as well about nationalism and diversity of opinions in almost every aspect of life. Global complexity appears at the moment to exceed humanity's capacity to rationalize problems and solutions, to identify global priorities and implement new policies for a common better, future.

Due to the explosion of the financial crisis in 2007 and further events until 2011, our society appears to be in the middle of the river, at the same time inevitably affected by the urgent need to identify structural changes in economy, implement new policies and take care of a survival which is being differently felt by nations in social and economic terms.

The way globalization has affected global trends in every sector of the human development depicts at the same time (i) the hope for a common affluent future of nations, (ii) the shareable aspiration to freedom of several countries still under dictatorial regimes and internal turmoil, (iii) the ambition of some communities to reach full independence and become nations and, (iv) the tragedy of poor nations. New critical issues qualifying the current transition are:

- financial markets are the new source of an impersonal and power which is reshaping the global economy

- a new order is at work at national and international levels. Profit dominates the economy and financial markets are the mechanisms

- the transfer of power from governments to financial institutions and the banking system is evident from the capacity of investors and the programmes adopted in the Business Schools and Universities worldwide. A number of Governments adapted to this new order, thus reducing their actions to a mere quest for money through new taxes and the absence of development programmes.

Globalization — evidently unavoidable as a consequence of scientific progress, population and GDP growth of Nations since 1990s — has heavily affected (i) the organization and the sovereignty of States, (ii) democracy, as represented by Governments selected through free elections, and, (iii) transnational corporations which have acquired an unprecedented power.

Under these conditions, the free-market economy is no longer controlled by government authorities — which in democracy are supposed to operate in collective welfare terms — rather by market operators and the banking system. As it usually happens during transitions, old rules fade away and new rules and procedures are pragmatically adopted.

The effect of globalization was to compress a variety of issues2[154] in a short time span, entirely revolutionizing a World until 1990 under the control of USA and

---

[154] Economy, population, environment and the ecosystem, ideologies, religious beliefs, lifestyles, culture and science, and interconnections through the web.

URSS. Globalization definitely affected the traditional structure and Governance of Nations rising the need for democracy.

The current condition of our society and different opinions and options available, are at the moment in contrast with the need for a globally sustainable development process, more easily achievable in the presence of an efficient democracy, in which equity and social justice prevail over private, personal and corporate interests.

Figuring out the future, whatever will be the solution for the environmental and economic crisis, is a formidable task which should be primarily centred on knowledge, awareness and participation.

The Cultural Revolution

The importance of education and awareness was synthesized in *The Global Partnership for Environment and Development, A Guide to Agenda 21*[155] as follows:

Education, public awareness and training should be recognized as a process by which human beings and societies reach their fullest potential. Education is critical for promoting sustainable development and improving the capacity of people to address environment and development issues it is also crucial for achieving environmental and ethical awareness, values and attitudes, skills and behaviour consistent with sustainable development and for effective public participation in decision-making.

The "cultural revolution" is needed on the basis of some unequivocal problems:

- in a few years the Earth will be increasingly overpopulated, overpolluted and more constrained by natural limits,

- the human imprint has kept growing faster after World War II, exceeding in the 1970s the biocapacity of the ecosystem and making our society unsustainable and heavily indebted with future generations

- the main challenge of our globalized society is to find a suitable balance between free economic competition, rules and key principles for a sustainable democratic society.

---

155   The Report *The Global Partnership for Environment and Development* was published in New York, (1993). The sentence is

The idea that large nations should be the first to start considering the RTM (Chapter 9) as a viable programme does not represent a discrimination against small nations; rather it is the recognition that time is running out and the implementation of the three sub-components (energy revolution, the sustainable development of infrastructure and the cultural revolution) needs great investments and rapid action.

Understanding the complexity of the environment, the dimension and dynamics of the global crisis, and adopting key principles for a common coexistence is a difficult but achievable cultural task. Educational programmes, from the primary school level onwards, should guide humanity along the pathway to the future. In the light of its inherent complexity, the cultural revolution should by carried out by:

- government authorities by reforming the educational sector and updating University programmes. Ecological Economics, Environmental Sciences and Welfare for a sustainable World should be considered as an indispensable common denominator for a variety of disciplines directly or indirectly related to human survival

- the people, who can today interact through internet, as a basic tool to interchange information, disseminate culture and make their voices heard at the national and international levels

- the society which is currently based on inadequate government policies, inefficient mechanisms of production and basically the idea that climate change is a tactic used by environmentalist to scare people.

Contradictions within the economic, financial and social system are many, but some are worse than others:

- a private sector based on free-market entrepreneurship in the absence of laws and controls is in contrast with the concept of equal rights and duties in democratic systems. The myth of an ever-growing production supported by population growth and increasing consumption (Figure 6.1), is inconsistent. The spiral-based process cannot go on indefinitely!

- the huge power of financial capital overwhelms the industrial capital and the capacity of governments to control domestic production. Investors, rather than governmental institutions, allocate funds, while consumer spending is the new market indicator

- emerging economies (now including nearly half the World's population) are providing low-cost labour, adequate industrial organization and higher profits, thus practically dividing the society into cheap producers and rich, but declining, consumers. Rich nations are simply unsustainable and this condition inevitably generates the loss of jobs and lower salaries, even in the case of a GDP growth.

- new technologies and automation create new jobs, but not enough to reduce unemployment. Either long-term labour-intensive strategies are adopted, or the growing population will inevitably succumb — even in emerging nations — to inescapable social and environmental limits.

- the possibility to save lives thanks to new discoveries (in biology, genetics and medical technologies) is in open contradiction with the huge investments in armaments. Military actions should be mainly limited to defensive and peace-keeping operations.

- population explosion is by far the most powerful driver of environmental impacts and resources overconsumption. If no limits to stabilize the birth rate are internationally adopted, then population will almost certainly decline due to starvation and low salaries.

The cultural revolution implies a reorganization of the human society which should update the old fashioned view of isolated nations in a permanent conflict in the economy and resource domains. Globalization most probably is an irreversible process that needs, however, to be supported by adequate policies and rules, should humanity start a new historical phase of progress.

The adoption of the SDP and RTM provides the framework for an orderly arrangement of the transition, more or less similar in concept to the process which went on in 1950s in the nations which had lost World War II. These countries carried out the reconstruction while changing the political system, introducing new democratic rules, implementing social changes among which free-press and an efficient free-market economy. The pathway to a sustainable society obviously takes much longer time today, by involving nations with differences in resources, technologies, political, cultural and social institutions. The transition can be conceived as a sequence of stages within the process of development, with the richest nations starting first. Involved countries should primarily concentrate on the current economic crisis, climate change, food for a growing population and the recognition of limits beyond which critical conditions are inevitable.

A wild globalization, as it is going on now, is in contrast with the stability of the international economic system. The cultural revolution should in sum take care of environmental issues and human limits which can definitely hamper shared solutions for a sustainable future.

The Waning of Humaneness

The decline of man is described in the book *"The Waning of Humaneness"*, published in 1983 by the Austrian naturalist and Nobel Prize Konrad Lorenz. According to Lorenz the technocratic system (which is composed of people with different technical expertise in market activities), rules government initiatives even in advanced democracies. Technocrats do not care about human values, individuals, the equitable distribution of resources, therefore they use ideologies and religion for economic purposes. The decline of humanity is therefore considered a common denominator of technocratic societies.

On the other hand the most advanced industrial plants, which are currently dimensioned on the needs of big nations, require huge standardized production (food, cars, house-hold appliances etc.) based on the national and international market competition.

The city-state of Athens at the time of Pericles (four centuries B.C.) carried the ideal dimension for the people to walk in the street, debate on ethics, logics and the ideal structure for their democracy. Population in 1983, when *"The Waning of Humaneness"* was published, amounted to 4.5 billion, compared to the 7 billion people in 2011.

In the last 30 years, our technocratic society underwent an explosive development, production activities became growingly international and interconnected, but also more vulnerable, thus leading the system towards a dangerous decline.

The utopia of a sustainable development process, despite the inherent difficulties in changing our complex society, represents a fascinating challenge. Reason, culture, sound key principles, science and the new technologies can transform this challenge into a real chance for humanity to survive inside the only cosmic ecosystem available. Pessimism in our current condition would definitely appear an inconclusive luxury!

CHAPTER 13

A Global Perspective for a Unilateral Action of Nations and a Proposal to The European Union

Introduction

The global perspective depicted in Figure 9.1 portrays the State of the World in 2004, and the possibility that the United Nations accept to be involved in a long-term Sustainable Development Process (SDP), while some nations jointly and unilaterally could advance towards a carbon-free society by adopting the Reference Transition Model (RTM).

Time is now for humanity to make a choice, bearing in mind that the international debate is dominated by the four major issues which could cause the society to collapse under its own weight: resources depletion, climate change, environmental degradation and the current financial-economic crisis.

The adoption of the RTM implies the involvement of nations the size of the USA, Russia, China, India and the European Union. These nations can be forced to initiate the process of change under the urgency of environmental and survival problems. Figure 10.3 shows the *adv E/R scenario 2010* by EREC & Greenpeace and gains which can be achieved through efficiency and the transition to renewables. The *adv E/R Report 2010* describes specific programmes for each of the "ten regions" identified by the IEA

OECD North America, Latin America, OECD Europe, Africa, Middle East, India, China, Developing Asia, Transition Economies, OECD Pacific,

on the basis of the following components:

- environmental crisis

- sustainable development

- climate, energy policy and security

- efficiency based on advanced technical solutions, new technologies associated with renewables and developments in the transport sector

- investments, expected results at the global and regional levels

- costs and benefits

- advanced cooperation between private and public sectors.

For each of the "ten regions" are developed different individual scenarios which reflect the variety of local conditions. Figure 10.5 shows the "OECD E/R Europe scenario", the Report providing data for the transition to a sustainable carbon-free world, through adequate solutions compatible with population, climate and resources. Even more specific is the tailor-made Report "*energy (r)evolution. Towards a fully renewable energy supply in the EU 27*".

Due to the current emergency some EU nations have already developed a broad national plan to solve energy problems by fostering the use of renewables. It is self-evident by now that the reduction of CO_2 emissions and the control of global warming are pre-requisites to limit the risk of climate change, and that solutions need to be identified and implemented within this decade. Detailed papers by EREC (as *Roadmap 2020* and *Targets 2020*) confirm the EU's commitment to increase the supply of energy from renewable sources.

The RTM Proposal to the EU Compared with other Regional-size Nations

The proposal to the EU to adopt the RTM (described in Chapters 9 to 12) does not exclude a similar choice by other Nations worldwide. China, India, Russia and the USA bear the topmost potential for a unilateral initiation of the RTM.

China and India, however, at present enjoy high GNP growth rates, as a result of their low-cost labour and technological capabilities, and are therefore more focused on their own political and economic agenda. Russia is endowed with enormous non-renewable sources (oil and gas), holds the greatest one-nation territory, is a powerful exporter of fossil fuels and the third emerging global power. It is less likely that these big nations will get involved in a proposal that at the moment for different reasons does not entirely match their pathways. These Nations could (and probably will) adopt selected solutions to minimize future environmental impacts.

The USA remains in the end the most advanced society that needs to reduce as soon as possible carbon emissions, implement the transition to renewables, lower the pollution level and reduce the impact of natural disasters.

Thanks to the contribution to environmentalism and the development of new and advanced technologies, USA can make structural changes before the further deterioration of the environment reshapes the global economy and the entire society. Whatever the choice made by nations, it remains central to the international debate to keep the planet habitable by preserving the environment, halting climate change and moving from growth to sustainability.

The EU, despite current problems, is the most advanced in conceiving the need for a carbon-free society, thus carrying the potential for a high level coordinated effort, by adopting the RTM (or a similar programme) and initiating a new era of advanced cooperation, among Union Members and other Nations, towards an ecologically sustainable development.

According to a sentence by the Sheikh Zaki Yamany[156], *"the Stone Age did not end for lack of stone"*. Similarly, the fossil fuels age does not have to cease for lack of oil, gas and coal, but mainly to halt the greenhouse effect, preserve the environment and the ecosystem and bring the climate change under control.

The Global and Regional Problems of the EU

The European Union emerged as a powerful idea from the devastation of World War II, and the founding fathers, aware of past difficulties, followed the dream of a Union based on ideals and democracy. Since the 1950s, the World has changed dramatically and the successful trend initiated by the EU, with the introduction of the Euro in 2002, was slowed down by the 2007 global financial crisis, the consequences of which are still affecting its economy. Some major issues dominate the EU internal debate.

The first involves the World's population, which amounts now to 7 billion people, and related income (National Geographic, March 2011 issue):

- 1 billion with a high income level of $12,000 or more per year

- 1 billion with an upper-middle income between $3,496 and $12,195/y

- 4 billion with a lower-middle income between $996 and $3,945/y

- 1 billion with a low income below $995 per year

The global imbalance is even more dramatic if the distribution of wealth is considered. The gap between rich and poor Countries is widening and while there is no limit to becoming richer, the extreme poverty of hundreds of millions of people in the world — accepted with silent indifference — has a limit: death from starvation. Europe is just across the Mediterranean and unless the condition of people in Africa, the Middle East and part of South Asia improves, the current massive migration will go on for decades.

The second concerns the condition of the environment as a consequence of global GHGs emissions, pollution and the effects of natural disasters. European

---

156 Sheikh Yamani was the Head of OPEC for many years and his original sentence is reported in the EREC/Greenpeace Report energy (r)revolution 2010.

Authorities should implement a new legislation on environmental protection, unifying rules and procedures into a shared sustainable development programme.

The third involves EU fuel import costs, which reached € 350 billion in 2008; in case of delay in the implementation of new policies, oil imports could rise further, hampering the development of renewables and lowering the amount of financial resources for clean technologies.

The fourth component is the successful performance of emerging economies (BRICS), at the moment taking advantage of lower labour costs and technological capabilities. The export imbalance between BRICS and the rest of the World may last for years, since new countries will join emerging economies.

In the absence of a radical change, EU competitiveness, with the exception of a few Member States, will decrease, unemployment rise and the Union be affected by an uncertain future.

The Proposal to the EU

The proposal is an invitation to the EU to consider the Reference Transition Model (RTM) as a multidisciplinary option involving scientific, technological, political, economic and cultural issues.

Part of the actions foreseen in the RTM have already been adopted unilaterally: the EU has pledged to a 20% reduction in CO_2 emissions, a 20% reduction in fossil fuel consumption through efficiency compared to the traditional business-as-usual approach, and to 20% more renewable carbon-free sources within 2020, compared to 1990 levels.

According to Sven Teske[157] from Greenpeace, the energy revolution is already underway at the global level. The European Photovoltaic Industry Association issued a report stating that the global PV capacity could increase from 36 GW in 2010 to 180 GW by 2015, and 350 GW by 2020, a capacity nearly ten times greater in ten years.

The last year's report by the Global Wind Energy Council stated that wind could support 12% of global demand by 2020, and 22% by 2030. Teske concludes quoting information from the Pew Charitable Trust according to which in 2009 China's clean energy investments amounted to $34.6 billion, compared with $18.6 billion by the US.

---

157 Sven Teske is the Director of the Renewable Energy Campaign for Greenpeace International. The energy [r]evolution has begun

The pathway towards a carbon free-society involves huge investments that can boost the European economy by keeping huge financial resources in the Union and creating more jobs.

The RTM initiatives *"energy (r)evolution", "sustainable development of infrastructures"* and *"cultural revolution"* are complementary interacting components for a sustainable future and carry the potential to open the way for a higher level of cooperation among countries. Tangible results in safe and clean energy, efficiency, market competition and employment could be achieved during this decade.

Conditions that qualify the EU for a stronger unilateral action are:

- a high cultural, scientific and technological level

- the stable size of the population

- the availability of financial resources and the entrepreneurial attitude of the people

- internal and external markets, which in a few more years could be eroded by the cheaper products from emerging nations, the rising cost of fossil fuels and a growing unemployment if the transition to renewables and energy efficiency proceeds slowly

- a relatively sober lifestyle and the cautious consumption of resources

- the possibility to initiate a sustainability process that would attract other countries, opening the way for a culturally advanced, democratic and free-market based future.

Compelling conditions for change are:

- dependence on imported energy. In 2005, the EU imported 82% of oil and 57% of gas (therefore being the World's leading importer of these fuels), and 97% of uranium. In 2007, Russia, Canada, Australia and Nigeria supplied to the EU more than 75% of its radioactive material

- the current employment, which could remain stable at the 2010 level or grow in a few member countries, but decline in others with effects on the entire Union

- the fact that EU investors are still tempted by emerging markets where labour costs are lower. Greater efficiency, lower costs of energy, new technologies and the recognition that current EU lifestyle can be preserved and even improved, can attract again private sector and foreign investors.

A guideline applicable to the EU was published in 2010 by *EREC and Greenpeace* under the title *"Energy (r)evolution", Towards a fully renewable energy supply in the EU 27"*. The report provides an overview of the initiatives that should be undertaken, new technologies, innovation in planning and the decentralization of energy sources, costs, savings and growing job opportunities. A number of publications concerning national, regional, global reports on the *energy (r)evolution* approach are provided by EREC and Greenpeace[158]

The responsibility of developing new technologies, within *adv energy (r)evolution 2010*, should not be left to market forces alone, but also involve Governments in sharing the financial risk: a joint effort could produce sounder initiatives and faster results. To enhance the interaction between the *adv E/R 2010* and the *sustainable development of infrastructures*, an efficient coordination is needed.

Finally RTM represents an unprecedented attempt to make the public and private sectors, cooperate closely and wisely as it should normally happen in case of emergency: there might be no other chances if humanity fails to curb CO_2 emissions and control global warming.

Environmental Ethics and the Need for a Rapid Transition

The global crisis incorporates an issue almost ignored at the international level, that of an *environmental and human ethics*. A number of scientific papers and books suggest that science and the new technologies can minimize impacts and pollution, thus providing the mediation tool between environment deterioration and human activities.

In a mere time span of 60 years, human survival turned into the growing and endless utilization of resources in the absence of rules and obligations. The difference between nature and humans is that the former "behaves" on a long-term basis, according to natural laws and phenomena, while the latter created in six decades the

---

158 http://www.erec.org/media/publications.html; www.erec.org/media/pubblication/greenpeace-eu-energy; http://www.greenpeace.org

condition for a better human life, but at the expenses of ecosystems, environment and the stability of natural cycles.

Environmental ethics can be regarded as the set of rules and principles that can keep together humanity and the natural world in the long term. By irresponsibly damaging environment and ecosystem the rights of human beings are violated in a foolish process of destruction, evidently constrained by the physical limits of the Earth and the growthmania of our society.

In this view sustainable development has to be considered a cultural process since it implies a higher level of knowledge, free-thinking, the target of a stable coexistence between nature and man, and a population mature enough to engage in a rapid transition towards a carbon-free society. Moral judgment should guide humanity in overcoming the global crisis by respecting the biosphere's uniqueness, bearing in mind that natural phenomena are the result of combined forces, some of which far beyond our control.

Another problem is represented by the food production which is likely to reach its upper limit in the current decade or so. In a context of growing population and food demand in poor countries and emerging economies, market prices inevitably rise uncontrollably. Food scarcity is currently affecting one billion people and the situation is likely to worsen in the years to come. The 2010–2011 revolt in North African nations is no longer a premonitory sign, but rather a dramatic outcome which makes people to risk and lose their lives in the attempt to reach countries that can ensure better survival conditions.

Never before have casualties been so high in the southern Mediterranean Sea, never so high was the percentage of migrants who reached Europe and the component which was repatriated. Our indifference to the problems of close-by Nations coincidentally disappears when oil and gas rich countries are involved! Ethics and a rapid transition to a sustainable world are essential to the initiation of a new man-nature relationship. As stated by the American astrophysicist Carl Sagan[159] in the article "*Croesus and Cassandra*", in the book *Environment in Peril*:

> *Out of the environmental crises of our time should come, unless we are much more foolish than I think we are, a binding up of the nations and the generations, and the end of our long childhood.*

The current unethical behaviour of nations involves:

---

[159] Carl Sagan (1934-1996), was an American astrophysicist and cosmologist who promoted the Search for Extra-Terrestrial Intelligence (SETI), carried out by The Planetary Society. The book *Environment in Peril* was published in 1991 by the Smithsonian Institution Press, edited by Anthony B. Wolbarst, a physicist at the US Environmental Protection Agency, Washington, DC.

- the political pragmatism of rich countries which privileges resources' over-consumption and global pollution

- the uncontrolled power of the banking system which acted in the past 10 years or so as an independent and transnational power

- the inadequacy of legislation. Externalities are tolerated by the public sector as an undeclared but real subsidy to every sort of activities (and impacts) of which unscrupulous investors take advantage; the inaction towards polluters and tax evasion feed illegal activities

- massive investment in armaments, which has been going on since World War II (as already highlighted in 1987 by the Brundtland Report). Nations should rely more on defensive systems

- the birth control which is still considered a crime by some organizations, so blind as not to recognize the suffering associated with a growing population affected by poverty, famine and unemployment.

The deadline for preventing the catastrophic scenario 2015–2040 in Figure 9.1 is now. Similarly urgent is the cultural revolution to increase the awareness levels of the people in general, politicians and policy-makers in particular.

Knowledge, Awareness, Indignation and Participation

The current emergency requires knowledge and awareness, civil indignation and individual commitment to change within the framework of a critical evaluation of human achievements and limits.

While progress in science and technology is marked by a continuous self-supporting trend, changing our habits, adapting our lifestyles to the carrying capacity of the planet and making development sustainable, are results not easily achievable.

The addiction to growth is rooted on our traditional background of poverty and the adaptation to an unsafe and risky environment, which dominated the scene until the beginning of 1800s. The huge migration waves towards the Americas and other rich unexplored continents, were essentially supported by the hope for unlimited growth as a precondition for happiness.

The current environmental crisis and the 2011 financial-economic crisis, are signs that human organization is no longer adequate to the task. The permanent conflict with the environment and the inconsistency of the economic system are in open contradiction with our survival. We need a formidable change in our social organization to prevent further problems, but government authorities are not ready yet to develop programmes.

Knowledge and awareness are the cultural keys to understanding the global signs[160] of the crisis, the current scenario and the risks for the future generations. Survival is no longer an instinctual issue only, rather the result of knowledge, shared social choices and moral judgement.

Indignation concerns humanity's response to the innumerable organizational anomalies, distortions and contradictions of our society (use of fossil fuels, deforestation and global pollution). The need for equal civil rights and the deregulation always claimed by the capitalist system are in open contradiction: the coexistence of capitalism and democracy needs to be revised.

Participation and commitment imply transition to action in a variety of forms: protest, information dissemination through Internet and any other channels that can allow people to communicate. Civil society needs to increase interaction with the national institutions, which are supposed to take care of the social welfare.

Do nothing or do it slowly makes the situation to deteriorate further. EREC & Greenpeace have provided *energy (r)evolution* programmes for a number of Nations, integrated within each of the ten regions identified by IEA. The sequence of transitions (population, environment, resources, climate, lifestyles) within the major global transition, also represents an opportunity: if we do not halt climate change, we will then need, in a few decades, to adapt to it in the course of a sequence of uncontrollable disasters.

Keep challenging the Earth's biodiversity and the human survival, in this astonishing and superb planetary assemblage of environment and ecosystems is an irresponsible action. Europeans seem to have forgotten the worries that in the 1950s brought R. Shuman, K. Adenauer, J. Monnet, A. De Gasperi to take the first steps towards the edification of a solid common home.

---

160 Premonitory signs of the crisis — like the rise of mean temperature, the shifting of the subtropical limits towards poles, the migration of species, the spreading of diseases, coral bleaching, the loss of pluvial and boreal forests, ocean acidification and the accelerated melting of glaciers worldwide — are unquestionable facts. The novelty is the current financial economic crisis, the stagnation of the US and the EU and, to a lesser extent, problems within the emerging economies.

EU citizens should draw inspiration from the motives of the founding fathers and consider adopting a model such as the RTM, or a similar one, with the potential to carry the Union and other Nations into the future.

Conclusion

Signs from the Earth/Biosphere make the difference between our planet as it was before the industrial revolution and as it is now:

- signs of physical changes which affect the surface of the planet. As a consequence of population growth, CO_2 emissions and global warming, uncontrollable phenomena — as the accelerated ice melting, the collapse of ices shelves worldwide, the rising of sea level and the warming of oceans — are likely to threaten people, their activities and the life in terrestrial and oceanic environments

- signs of abrupt ecological changes are heavily involving ecosystems: habitats change, coral reefs bleach, exotic species migrate and the acidity of oceans increases

- signs of time: phenomena that previously occurred slowly — as for example climate changes associated with the sequence of glaciation and deglaciation stages during the last half a million years — are now taking place during a life span. Himalaya glaciers can disappear by 2035 or so, but many others worldwide, smaller in size and at lower altitude, are already vanishing. By 2050, less than 40 years from now, the polar regions can turn into nearly iceless places, the World becoming a hothouse, as it was at the time of PETM 56 million years ago

- signs of unprecedented cultural, social, economic and stress conditions. Our society is inadequate in understanding that growthmania is already beyond limits and on-going changes are innumerable, while several nations are still striving for the recognition of equal human rights and democracy. Corporate interests and misinformation dominate World markets.

Population growth and activities, the rapid depletion of resources and the huge emission of GHGs depict a species in trouble. People are slowly but inexorably, however, becoming aware that further progress can be hampered by the combined effects of climate change, the consumption of non renewable resources, food scarcity and widespread pollution.

The Author is aware that his proposal for a sustainable development may be considered both visionary and naïve. Nevertheless, the fact remains that the ecosystem is globally declining at a rate much faster than politicians, traditional economists and the majority of people can conceive. Our increasingly integrated world may be likened to an avalanche out of control, rolling down a slope, while taking up speed and growing in size, indifferent to natural obstacles and limits, and destined to end up as a chaotic mass of debris.

We can halt the trend by modifying our mind-set, learning about our world and its dynamics, and most importantly, by changing our habits and adopting a more responsible behaviour.

Very slowly over more than nine millennia, and then, through a rapid upheaval since the industrial revolution, humanity exploded putting at risk the stability of the biosphere and its capacity to support life in general and our species in particular.

No one with any sense would organize a weekend in the mountains, using an old camper in bad conditions, overloading it with people and baggage, being at the same time aware that brakes are inadequate to the weight and the speed of the vehicle. Nevertheless, this is now happening with the recent evolution of humanity on Earth, a journey that can end with a global catastrophe.

Above all, the "growth paradigm" is inherently in contrast with "sustainable development", which relies on the dynamic equilibrium between the demand for resources and the natural regeneration rate and absorption/recycling of waste.

Under these conditions, the transition towards a sustainable world during the 21st Century is conceivable as the only way to support current and future generations, within a framework of equal, or nearly equal, opportunities. It is in fact clear that non-renewable resources are being rapidly depleted.

The need for a sustainable development was already felt as a big problem in 1987, when the Bundtland Report was published. The evident difficulty was in providing an *operational methodology* for estimating human activities and needs, and the pathway for a smooth transition towards a sustainable development based on rules and procedures.

The RTM is the operational approach which accounts for the current complexity of the system, takes into consideration priorities, limits (natural and human) and key principles, and provides feasible solutions.

Advantages and problems, related to a controlled transition towards a sustainable development, need to be compared with impacts associated with human overpopulation, the current production system and the growth economy. The fear of a long-term sustainability process, as the enemy of the free market, is undermining the progress of democracy worldwide and widening the gap between rich and poor people. A deregulated free trade in a congested and overpolluted world of growing inequality and global risk, is neither acceptable nor useful: human, social and environmental costs are growingly greater than benefits.

References

Chapter 1

J. X. Kasperson & R.E. Kasperson - Global Environmental Risk (2001), The United Nations University

B. Commoner - The closing circle (1971), Alfred A. Knopf Inc

S. Mines - The last days of mankind. Ecological Survival or Extinction, (1971). Simon and Schuster, N.Y.

B. Ward and R. Dubos - Only one Earth (1972), Andre Deutsch ltd

B. Commoner - The poverty of power (1976), Bantam Books

J. Lovelock - Gaia: A New Look at Life on Earth (1981) Oxford University Press

J. R. Taylor - The Doomsday book (1972), Book Club Associates

M. Wackernagel & W. Rees - Our ecological footprint (1996), The New Catalist

J. Lovelock - The revenge of Gaia, (2006), Penguin Books

R. Carson - The Silent Spring (1962), Houghton Mifflin

Chapter 2

F. Hoyle - The Origin of the Universe and the Origin of Religion (1993), The Frick Collection

M. Rees - Our Cosmic Habitat (2001), Princeton University Press

J. D. Barrow - The constants of Nature. From alpha to omega (2002), Pantheon Books N

M. Rees - Our Final Hour (2004), Basic Books N.Y

S. Hawking & R. Penrose - The Nature of Space and Time (1996) Princeton University Press

Chapter 3

J. D. Macdougall - A short History of Planet Earth (1996), John Wyley & Sons

R. Leakey & R. Lewin, The Sixth Extinction. The pattern of Life and the Future of Humankind (1995) Doubleday New York, London, Toronto, Sydney, Auckland

W. Ryan & W. Pitman - Noah's Flood. The new scientific discoveries on the event that changed history (1998), Simon and Schuster N.Y:

Y. Baskin - The work of nature. How the diversity of life sustains us, (1997), Island Press

T. Volk - Gaia's Body. Towards a physiology of the Earth (2001)

D. M. Raup - Extinctions. Bad Genes or Bad Luck? (1991), W. W. Norton & Company

Chapter 4

R. E. Leakey - The Making of Mankind (1981), The Rainbird Publishing Group Ltd.

A. Lommel - Prehistoric and Primitive Man (1966), The Hamlyn Publishing Group Ltd.

A. Peccei - Quale Futuro, Mondadori (1974)

J. Shreeve - The Greatest Journey, The trail of our DNA (March 2006), National Geographic,

Chapter 5

IPCC - Climate Change Report (2007), AR4

P. Hoffmann - Tomorrow's Energy, Hydrogen, fuel cells and the prospect for a cleaner Planet (2001), MIT Press

J. X. Kasperson & R.E. Kasperson - Global Environmental Risk (2001), The United Nations University

Global Warning, Bulletins from a Warmer World (September 2004) National Geographic

M. Bell & M.J.C. Walker - Late Quaternary Environmental Change (2005) Prentice Hall

W.F. Ruddiman, T.J. - The case for human causes for increased atmospheric CH4 over the last 5000 years (2001), Quaternary Science Review

Chapter 6

C. H. Mooney - The Republican War on Science (2005), Basic Books

R. B. Reich - Supercapitalism. The transformation of Business, Democracy and Everyday Life(2007) Alfred A. Knopf, New York

J. D. Sacks - Common Wealth, Penguin ,2009

J. E. Stiglitz and L.J. Bilmes,- The Three Trillion Dollar War, 2008, W.W. Norton

N. Klein - The Shock Doctrine, 2007, Klein Lewis Production Ltd

J. K. Galbraith - The Great Crush, 1954, Houghton Mifflin, Boston

J. K. Galbraith - Economics and the public purpose, 1973, Houghton Mifflin, Boston

REFERENCES

J.K. Galbraith - A short history of financial euphoria, 1994, Penguin Group

J.K. Galbraith - The New Industrial State, 1967, Houghton Mifflin, Boston

J.K. Galbraith - The economics of innocent fraud (2004), Houghton Mifflin, Boston

H. Marcuse - One-Dimensional Man. Studies in the Ideology of Advanced Industrial Society (1964), Beacon Press, Boston

H. Marcuse - Saggio sulla Liberazione (1969), Giulio Einaudi Editore

Chapter 7

The Brundtlandt Report (1987), United Nations

H. E. Daly - Beyond Growth (1996), Beacon Press

J. Farley, J. D. Erickson, H. E. Daly - Ecological Economics. A workbook for Problem-based Learning (2005), Island Press

Chapter 8

Millennium Ecosystem Assessment, 2005, UNDP

J. W. Forrester - World Dynamics, Wright-Allen Press, Inc., 1971

J.W. Forrester, D.L. Madows, J. Randers, A.A, Anderson, J.M. Anderson, W.W. Behrens

III, R.F. Nail, S.B. Shantzis - Toward Global Equlibrium, 1973, Wrigth-Allen Press, Cambrindge Mass.

D. Meadows et al. - Limits to growth (1972), The Sustainability Institute

D. Meadows et al - Beyond the limits (1992), Chelsea Green

D. Meadows et al. - The New limits to growth: the 30-years update (2004), Chelsea Green

M. Wackernagel & W. Rees - Our Ecological Footprint (1996), New Society Publishers

Living Planet Report. (2010), WWF

Stokstad E - Taking the pulse of Earth's life support systems, Science (April 1 2005), Science

Chapter 9

J. ROCKSTRÖM - The Stockholm Resilience Centre (Sweden)

J. Hansen et al. - Target Atmospheric CO_2: Where Should Humanity Aim?

National Geographic, June 2004 - The end of Cheap Oil

J. Farley, J. D. Erickson, H. Daly - Ecological Economics. A workbook for Problem-Based Learning (2005), Island Press

M. Wackernagel & W. Rees - Our ecological footprint (1996), New Society Publishers

Chapter 10

Energy (r)evolution, 2010. A Sustainable World Energy Outlook, EREC/GreenPeace

World Energy Outlook (2010), International Enery Agency

J. Schindler et al. - Crude Oil; The Supply outlook (October 2007), Energy Watch Group

Global Warning, Bulletins from a Warmer World,(September 2004), National Geographic

Earth's Future, Solutions for a finite world, Special Report (April 2010), Scientific American

M.Z. Jacobson & M.A. Delucchi - A Plan for a Sustainable Future (November 2009), Scientific American

Policy Guide on Planning for Sustainability (2000), American Planning Association (APA)

M. Kenny & J. Meadowcroft - Planning Sustainability (2002), Taylor & Francis Inc

S.M. Wheeler - Planning for Sustainability (2004), American Planning Association

Chapter 12

P. Ehlrich - Can we respond to the growing environmental threat to civilization? from Environment in Peril (1991), Smithsonian Institution Press

C. Sagan - Croesus and Cassandra. Policy Response to Global Change, from Environment in Peril, (1991), Smithsonian Institution Press

The Millennium Development Goals 2010, UNDP

D. Coyle - The Economics of Enough: How to Run the Economy as if the Future Matters (2011), Princeton University Press

Chapter 13

Robert Kunzig - Population 7 Billion, (January 2011), National Geographic

B. Commoner - The closing circle (1971), Bantam Book

Karl Kraus - Last days of mankind (1974), Ungar Pub. Co.

Ward and R. Dubos - Only one Earth (1972), United Nations

Links

Mass Extinction Underway (Chapter 1):
    http://www.massextinction.net/
Stockholm Resilience Centre (Chapter 1):
    http://www.stockholmresilience.org/planetary-boundaries
WRI - Millennium Ecosystem Assessment, 2005 (Chapter 1):
    http://www.wri.org/publication/millennium-ecosystem-assessment
Mendeleev Table (Chapter 2):
    http://abyss.uoregon.edu/~js/ast123/lectures/lec21.html
Periodic Table of the Elements (Chapter 2):
    http://www.mendeleevtable.com/
Adam Dimech Online - Continental Drift Maps (Chapter 3):
    http://www.adonline.id.au/plantevol/maps/
NOAA Ocean Explorer (Chapter 3):
    http://oceanexplorer.noaa.gov/explorations/05fire/background/volcanism/media/tectonics_world_map.html
USGS - Understanding plate motions (Chapter 3):
    http://pubs.usgs.gov/gip/dynamic/understanding.html
Great Oxygenation Event (Wikipedia) (Chapter 3):
    http://en.wikipedia.org/wiki/Oxygen_Catastrophe
The Planetary Society (Chapter 3):
    http://www.planetary.org/
Global Warming: East-West Connections (Chapter 5):
    http://www.columbia.edu/~jeh1/2007/EastWest_20070925.pdf
Paleoclimate Implications for Human-Made Climate Change (Chapter 5):
    http://www.columbia.edu/~jeh1/mailings/2011/20110118_MilankovicPaper.pdf
The Current Major Interglacial (Chapter 5):
    http://www.roperld.com/science/currentmajorinterglacial.pdf
NWS JetStream - The Earth-Atmosphere Energy Balance (Chapter 5):
    http://www.srh.noaa.gov/jetstream/atmos/energy.htm

NSIDC Press Room 2009 Arctic sea ice minimum (Chapter 5):
http://nsidc.org/news/press/20091005_minimumpr.html
Models Underestimate Loss of Arctic Sea Ice (Chapter 5):
http://nsidc.org/news/press/20070430_StroeveGRL.html
Satellite Gravity Measurements Confirm Accelerated Melting of Greenland Ice Sheet (Chapter 5):
http://www.sciencemag.org/cgi/content/abstract/313/5795/1958
Bad Sign For Global Warming: Thawing Permafrost Holds Vast Carbon Poo (Chapter 5):
http://www.sciencedaily.com/releases/2008/09/080903134309.htm
The Greenhouse Effect And Climate Change (Chapter 5):
http://www.bom.gov.au/info/GreenhouseEffectAndClimateChange.pdf
Estimating World GDP, One Million B.C. – Present (Chapter 6):
http://econ161.berkeley.edu/TCEH/1998_Draft/World_GDP/Estimating_World_GDP.html
GDP; PPP (US dollar) in World (Trading Economics) (Chapter 6):
http://tradingeconomics.com/world/gdp-ppp-us-dollar-wb-data.html
Economy of the European Union (Wikipedia) (Chapter 6):
http://en.wikipedia.org/wiki/Economy_of_the_European_Union
List of countries by past and future GDP (PPP) (Wikipedia) (Chapter 6):
http://en.wikipedia.org/wiki/List_of_countries_by_past_and_future_GDP_(PPP)
The Irrationality of Homo Economicus (Chapter 7):
http://www.iisd.org/didigest/special/daly.htm
Five policy recommendations for a sustainable economy (Chapter 7):
http://www.feasta.org/documents/feastareview/daly2.htm
WWF Living Planet Report 2010 (Chapter 8):
http://wwf.panda.org/about_our_earth/all_publications/
WWF Living Planet Report 2006 (Chapter 8):
http://wwf.panda.org/about_our_earth/all_publications/living_planet_report/living_planet_report_timeline/lp_2006/
Tipping towards the unknown (Chapter 9):
http://www.stockholmresilience.org/planetary-boundaries
Chelsea Green (Chapter 9):
http://www.chelseagreen.com/

Scripps CO_2 Program (Chapter 9):
  http://scrippsco2.ucsd.edu/
2007–2008 world food price crisis (Wikipedia) (Chapter 9):
  http://en.wikipedia.org/wiki/2007-2008_world_food_price_crisis
Jay Haston, *From Capitalism to democracy* (Chapter 9)
  http://dieoff.org/
Recent Mauna Loa CO_2 (Chapter 10):
  http://www.esrl.noaa.gov/gmd/ccgg/trends/co2_data_mlo.html
Renewable Energy (Wikipedia) Chapter 10):
  http://en.wikipedia.org/wiki/Renewable_energy
Energy Watch Group (Chapter 10):
  http://www.energywatchgroup.org/
Target Atmospheric CO_2: Where Should Humanity Aim? (Chapter 12):
  http://www.columbia.edu/~jeh1/2008/TargetCO2_20080407.pdf
The energy [r]evolution has begun (Chapter 13):
  http://www.grist.org/renewable-energy/
  2011-02-19-the-energy-revolution-has-begun
European Renewable Energy Council (Chapter 13):
  http://www.erec.org/media/publications.html
Greenpeace EU Energy [R]evolution scenario 2050 (Chapter 13):
  http://www.erec.org/media/publications/
  greenpeace-eu-energy-revolution-scenario-2050.html
Greenpeace International (Chapter 13):
  http://www.greenpeace.org/
Trends in natural disasters (Chapter 13):
  http://maps.grida.no/go/graphic/trends-in-natural-disasters
Tonne of oil equivalent (Wikipedia) (Chapter 13):
  http://en.wikipedia.org/wiki/Tonne_of_oil_equivalent

Index

Adv E/R 2010, EREC/GREENPEACE, 150, 158
Albedo, 57, 64, 66, 78, 80
Altamira, 51
Anthropocene, 136
Ardipithecus, 46, 48
Australopithecus, 46, 48
Bethe H., 23
Big Bang, 12-24, 41, 44
Biofuels, 160
Biocapacity, 11, 121-130, 174, 178
Biodiversity, 2, 3, 5, 34-36, 60, 104, 114, 118, 123, 128, 136, 140, 192
Biosphere, 2, 3, 6, 7, 9, 11-17, 24, 27-34, 38-44, 54, 57, 60, 72-76, 98-108, 123, 130, 134-136, 149, 165-175, 190-194
Biosphere yearly energy budget, 72
Bipedalism, 46
Brundtland Report, 10, 95-101, 134, 140, 191
Burges shales, 40
Cap and trade, 151
Carbon Capture Sequestration, 161
Capitalism, 11, 83-94, 111, 141, 142, 176, 192
Carrying capacity, 103, 121, 123, 130, 191
Catarrhine monkeys, 46
Cave paintings, 51
Cenozoic Era, 34, 59-61
Chapelle aux Saints, 51
China, 27, 48, 49, 86-93, 111, 123, 143, 154-165, 184-187
Clausius R., 86, 89
Climate change, 2-13, 34, 39, 48, 53-72, 76-81, 96, 103, 111, 120, 128, 135, 136, 149-166, 175-186, 192-194

Clorofluorocarbons (CFCs), 76, 78, 79
Club of Rome, 113, 116, 117
Coal, 34, 49, 51, 57, 89, 137, 146, 153- 165, 173, 186
Communism, 83, 90, 175
Daly H., 90, 99-110, 131, 141,
Dansgaard-Oeschger events, 69
Democracy, 11, 83-85, 89, 92-94, 111, 141, 142, 176-178, 181, 186, 192-195,
Desertification, 4, 63, 98, 128
Deuterium, 24, 56-62
DNA, 39, 43, 51
Dryas, 71, 80, 135
Earth, 2-17, 24-46, 53-78, 84, 92-103, 111, 116, 121-123, 129-135, 146, 157-178, 190-194
Earth's magnetic field, 29, 37, 38
Ecological economics, 99, 102, 131, 132, 179
Ecological Footprint Analysis, 121, 126, 130
Economy 8-12, 72, 81-111, 118, 120, 129-143, 157-165, 176-188, 195
Ediacaran fauna, 36, 40
Eemian, 50, 70, 71, 79
Energy balance diagram, 72
Energy sources, 4, 145-153, 158, 165, 189
Eukaryota, 39, 40
Externalities, 90, 91, 108, 109, 141-153, 165, 191
Factors of climate change, 12, 54, 57, 58
Fossil fuels, 8-12, 51, 64, 65, 76, 81, 98, 104-108, 140-159, 165, 185-192
Fukushima disaster, 166
Galbraith, J. K., 9, 84, 90, 93
Geological time scale, 34
Globalization 93, 94, 99, 109, 110, 117, 176-181
GDP (PPP), 52, 86, 88, 92-98, 153, 154, 176-180
GNP and SSNNP, 109, 141
GOE, 36
Greenhouse effect, 29, 58, 65, 72-80, 134, 186
Greenhouse gasses 68, 78, 161
Greenland, Vostok Station, 59
Hansen J., 64, 136, 139, 173,
Homo Sapiens, 12, 15, 36, 45-50, 69, 70

INDEX

Homeostasis, 38
Hoyle F., 17, 21, 24, 25
Ice ages, 12, 62, 64, 66, 68, 71, 76, 81
Ice cores, 58, 61, 62, 67, 69
IEA/WEO 2010 Scenarios, 158-165
Insolation, 53, 56
Klein N., 92
Lisiecki L.E. & Rymo M.E., 61
Malthus T., 86, 88
Mammals, 3, 15, 34-38, 43-46, 51, 64
Meadows D.H, Meadows D.L, Renders J., Beherens III W.W, 117
Mendeleev, 18, 21, 22
Milankovitch, 55
Mill, J. S., 89
Millennium Ecosystem Assessment, 6, 10, 97, 113-115
Nucleosynthesis, 18, 19, 23-25
Orò J., 42, 44
Pangea, 60
Peccei A., 52, 116, 120
Planetesimals, 28
Platyrrhine monkeys, 46
Prokaryota, 40
Proxy data, 58
Radiative forcing, 60, 72, 76, 77
Rees M., 17, 23, 25
Reference Transition Model, 12, 13, 133, 137-142, 184, 187
Reich R.B., 11, 92, 111
Rifkin J., 91
RNA, 39, 43
Ryan W. & Pitman W., 49
Sachs J., 92
Sagan C., 16, 190
Scenarios 12, 58, 68, 79, 113-119, 128-131, 147-152, 156, 159-151, 184
Smith A., 86-88, 174,
Solar Energy 29, 38, 39, 42, 53, 55, 65, 72, 73, 140, 141, 146, 149
Solar System Planets, 27
SSNNP, 109, 141

Stockholm Resilience Centre, 2, 135, 136
Supercapitalism, 115, 86, 92, 94, 111, 176
Sustainable development process, 12, 96, 111, 126, 132-142, 152, 178-183
Tetrapods, 36
Thermohaline circulation, 66, 69, 79, 80
Throughput, 102, 110
Wurm Ice Age, 48, 69, 70

Glossary

Acidification
A change in the environment's chemical balance due to the rising concentration of acid elements. Ocean acidification is caused by a decrease in the pH of sea water due to the uptake of anthropogenic CO_2 and has negative effects on the production of calcium carbonate shells of oceanic calcifying organisms.

Aerosols
Aerosols are small solid particles of liquid droplets suspended in a gas that can generate smog and behave as greenhouse gasses.

Albedo
Solar radiation reflected by the surface of the Earth. Ice and snow-covered areas have a high albedo level (reflectivity), compared to the low albedo of oceans and vegetation. When global warming heats up the Earth, ice covers begin melting and the overall albedo effect decreases, thus accelerating the absorption of solar radiation and making the mean surface temperature rise.

Anthropogenic impacts
Human (or anthropogenic) activities carry environmental impacts, among which pollution of the air, sea and land, and heavily affect the stability of climate.

Atmosphere
The gaseous shell surrounding our planet is composed of Nitrogen (78%), Oxygen (21%) and very small quantities of other gasses (among which Carbon Dioxide (CO_2) and Ozone (O_3).

Biodiversity
The diversity of living organisms in marine-oceanic, terrestrial and aquatic ecosystems.

Biocapacity
The term concerns the capacity of a biologically productive zone to provide renewable resources and absorb waste.

Biomass
It is the total mass of biological material (living or recently dead) and includes plants, wood, waste and gas. Wood has been used for millennia for heat and cooking and

more recently for the generation of electricity and the production of biofuels for combustion engines.

Calving
Calving is a process of disruption of ice bodies involving the generation of cracks and the disintegration of mass. Greenland's glaciers ending in oceanic waters due to calving break up yearly into thousands of icebergs.

Cap and trade
In the attempt to reduce global warming the 1997 Kyoto Protocol fixed GHG emission targets and timetables, establishing a cap and trade system, based on tradable pollution permits. The target was to limit (cap) the quantity of GHGs in the atmosphere by allowing companies which succeeded in reducing their emissions to sell permits to others with exceeding emissions on the basis of a compensation system. Criticism against the cap and trade system concerns the inadequacy of controls, its limited effect compared to the quantities of GHGs emissions and the financial support that many polluters were able to take advantage of.

Carbon Capture Sequestration or Carbon Capture Storage (CCS)
This process involves capturing CO_2 from fossil fuels-based power plants, storing it into natural deposits (like exhausted mines or geological rock formations) from which it cannot disperse into the atmosphere. The process is expensive and at the moment does not appear to be a feasible solution to reduce GHGs concentration in the atmosphere.

Carbon dioxide
CO_2 is a most important natural gas and plays a fundamental role in photosynthesis. Carbon dioxide is generated by the combustion of fossil fuels, biomass, the burning of forests, industrial processes and is also emitted during volcanic eruptions. Its excessive quantity in the atmosphere affects the Earth's radiative balance, therefore enhancing the greenhouse effect and global warming.

Carbon footprint
The term involves the amount of greenhouse gasses (GHGs) produced by human activities (generation of electricity, heating, transportation and, in general, by combustion engines) at a global, regional, national or local level. The carbon footprint includes all GHGs gasses and is expressed in tonnes of carbon dioxide equivalent (CO_2-eq).

Carbon sink
Carbon sinks are natural or artificial reservoirs that store carbon-containing substances. Carbon reservoirs are the atmosphere, the hydrosphere, sediments, carbonaceous rocks and living organisms (through photosynthesis). In response to the significant importance and the negative effects of excessive carbon dioxide

emissions, the 1997 Kyoto Protocol (ratified in 2010 by 191 countries) attempted to fight global warming by controlling gaseous emissions and stabilizing GHGs concentrations in the atmosphere.

Climate change
The UNFCCC (United Nations Framework Convention on Climate Change) attributes the current climate change directly or indirectly to human activities, which add to the natural variability of climate.

Coal
Coal is the most abundant fossil fuel on Earth, the largest non-renewable source of electricity generation and the biggest driver of carbon dioxide emissions. Its layers (coal seams) originate from the accumulation of plant matter at the bottom of water bodies subsequently protected by mud layers which prevent oxidation. During the Carboniferous period (359–299 million years ago), the Earth's continents were unified in a unique landmass (Pangea) and shallow sea water conditions were so common that repetitive cycles of coal-seam formation gave rise to huge deposits worldwide. Coal is concentrated 27% in the USA, 17% in Russia, 13% in China, 10% in India, 9% in Australia, 5% in South Africa and 19% in other countries. In sum, 6 countries hold in 2011 81% of coal reservoirs and nearly half of the global population. Out of 5,440 million tonnes of production in 2006, 1,531 were used in China, 1,117 in Europe, 1,094 in the USA, 431 in India, 251 in Russia and 1,016 in other countries (National Geographic, March 2006). Despite the potential CO_2 emissions associated with the use of coal, global consumption is expected to rise to 8,200 million tonnes by 2025.

Coral bleaching
The widespread bleaching of corals and the deterioration of coral reefs are considered as a sign of the fact that the climate threshold beyond the optimum climate condition at which the reproduction of these marine organisms takes place was exceeded.

Desertification
The progressive degradation and expansion of desert areas into surrounding belts is called desertification and is attributable both to the natural variations of climate and to human activities, that can influence it directly and indirectly. The fastest desertification process is currently taking place in Africa, but signs of it were already present at the time of the Roman Empire.

Drought
It is the prolonged lack of precipitations and the related hydrologic imbalance. The process can be part of natural cycles lasting years or decades, or be the beginning of a long-lasting process eventually turning into a progressive desertification of the area.

Ecosystem, ecosystem services
The ecosystem involves the complex of living organisms and the surrounding environment behaving as a functional unit of interacting components. The Millennium Ecosystem Assessment Report 2005 by the United Nations highlighted the importance of ecosystems and "ecosystem services" as factors stabilizing the biosphere by providing resources and enhancing the well being of life (see Chapter 1).

Energy radiation balance
The Earth's radiation balance refers to the amount of solar energy entering and leaving the planet (see Chapter 5) and is expressed by an equation with a number of variables. Climate change on the Earth is conditioned by a variety of interacting factors (astronomical, terrestrial and human), thus being subject to a permanent time-variable instability. The current state of the radiation balance is influenced by the anthropogenic emissions of GHGs, which are considered responsible for global warming and the current climate trend.

Erosion
A most important natural process involving a variety of chemical and physical phenomena which in the short term result in the removal of sediments and in transportation by the river network. In the long term, erosion dismantles mountain ranges, being a most incessantly acting force of nature. In the long run, erosion concurrently flattens reliefs and creates submarine deposits which as a result of tectonics can be uplifted far above the sea level.

Forces of nature
Four forces described in physics control, in decreasing strength, natural phenomena: strong interaction, electromagnetism, weak interaction and gravity. Gravity affects all bodies in the Universe (Newton's law); electromagnetism generates interactions among particles charged with electricity within the electromagnetic field; strong interaction is a nuclear force (which affects quarks and antiquarks) binding together protons and neutrons in the atomic nucleus, while weak nuclear force is associated with radioactive decay.

Global Warming
It is the increase of the average global temperature as a result of the greenhouse effect currently associated with human activities. It is worth recalling that average global temperature is the result of a complex procedure based on instrumental data which are not evenly distributed worldwide. Compared to pre-industrial times, the average temperature rise in 2011 is around 1°C, but local temperature variations, which are responsible for the accelerated melting of the Greenland ice cover, are within the seasonal range of 10 °C.

Greenhouse effect, greenhouse gas (GHG)

Major greenhouse gasses in the atmosphere are water vapour, carbon dioxide, methane, nitrous oxide, ozone and chlorofluorocarbons (CHFs). The influence of GHGs (a function of gas type and abundance) on the greenhouse effect is estimated between 36% and 72% for water vapour, 9% to 26% for CO_2, 4% to 9% for methane and 3% to 7% for ozone. Therefore, within the current global warming trend and the related growing evaporation, water vapour is supposed to influence climate significantly.

Halocarbons

Halocarbons include: HCFCs (hydrochlorofluorocarbons), CFCs (chlorofluorocarbons), HFCs (hydrofluorocarbons), with the potential to deplete the ozone layer and cause skin cancer and cataract.

Land ice in Antarctica and Greenland, the Arctic Ocean ice cover and the Tibetan plateau ice sheet

The largest land ice sheet masses (greater than 50,000 km^2) are in Antarctica and Greenland. Antarctica ice covers 14 million km^2, has a volume of 30 million km^3, represents at the moment around 90% of the freshwater on the planet's land and if it were to melt completely would increase the global sea level by 61 m. The Greenland ice sheet is much smaller (1.7 million km^2) and its total thaw could make the sea level rise by around 7 m. Its current melting rate per year is 239 km^3. During the coldest stage of the last glaciation (from 90 ky to 20 ky years ago) larger continental glaciers existed: the Laurentide in North America, the Weichselian in Northern Europe and the Patagonian ice sheets in South America. In 2007, the minimum extension of the sea-ice cover reached 4.52 million km^2 in the Arctic Ocean [161]. The rate of surface decline of the cover rose from 2.2% per decade in the period 1979-1996, to 10.1% in the decade 1998-2008. The thaw of the oceanic floating ice cover, drives the reduction in cover extension lowers the albedo effect, thus exposing an increasing sea water extension to sun radiation and the rise in heat absorption. With an area of 1.65 million km^2, the Tibetan Plateau is the third largest freshwater reservoir in Asia. Faster snow and ice thaw in the area, rises global warming, causing destructive floods in India and Bangladesh and lowers the Chinese freshwater reserves.

Interglacial stage

It is the warmer period of around 10,000 years between two glacial periods, which, during the last half a million years lasted on average 90,000 years.

Moraine

Glacial deposits of soil and rock debris, resulting from the melting of land glaciers.

Neolithic age

---

161   Source: Climate Change Science Compendium, UNEP, 2009 (Earth Systems).

The New Stone Age began around 10,000 years ago in the Middle East.

Nucleosynthesis

There are two types of nucleosynthesis: the first resulting in the formation of H, He and small quantities of Li and Be when the Universe is about 200 seconds old, and the second, inside giant stars, when the Universe is some billion years old, and star explosions generate the other elements (Mendeleev Table).

Ozone Hole

According to a NASA-led study an unprecedented depletion of the ozone layer five times larger than California has been discovered above the Arctic Ocean (data from Internet, October 4, 2011)

Radiative forcing

It is the change in net irradiance at the level of the troposphere and the stratosphere which alters the balance between incoming and outgoing solar radiation. That difference, (IPCC Climate Change Report 2007) amounts to 1.72 W/m², mostly due to CO_2 from fossil fuels.

Permafrost

It is top level soil that remains below the freezing point of soil for a number of consecutive years. Permafrost is mostly located in the Northern hemisphere, in emerged land around the Arctic Ocean. If permafrost melting due to rising temperatures were to occur, huge quantities of carbon contained in peat (decayed vegetation) and methane (the natural gas CH_4) would be released.

Photosynthesis

The process through which vegetation absorbs solar radiation and carbon dioxide, releasing oxygen and producing carbohydrates, thus initializing the food chain. The accelerated decline of pluvial and boreal forests therefore represents a dramatic threat to the dynamics of the ecosystem.

Thermal expansion

When oceanic waters warm up as a consequence of the global temperature rise, the thermal expansion in volume (and lower density) of waters takes place.

Supernova

The explosion of a very large star is called a supernova. The phenomenon takes place when a star contracts and then violently explodes, becoming so bright to be visible, if close to the Earth, even in day time. Heavy chemical elements, among which the bricks of life (C, O, N, S, P), are generated during the explosion of giant stars at extremely high temperature and pressure.

Thermohaline Circulation (THC)

The global, oceanic circulatory flow of marine water, termed Thermohaline Circulation, is driven by variations in sea water density, rising salinity or decreasing temperature (Figure 5.10). THC, also named the ocean conveyor belt, and in the Mid- Atlantic the Gulf Stream, carries to North Eastern European countries warm (and food-rich) waters which make Scandinavia and the British Islands warmer compared to Canada and Greenland. The shutdown of the THC can cause a global temperature inversion and the beginning of a new ice age.

Transitional economy

A transitional economy is an economy passing from a centrally planned to a free-market economy (mostly refers to Former Soviet Union countries).

Tsunami

Huge marine waves associated with underwater earthquakes, volcanic eruptions or landslides. The April 2011 Miyagi earthquake in Northern Japan (9.0-magnitude) was followed by an enormous tsunami with waves up to 37 m high which devastated the North Eastern coast and caused Fukushima disaster.

Wetland

Wetlands are composed of permanently or seasonally moisture-saturated soil including swamps and marshy areas, rich in biodiversity. Through the Ramsar Convention (1971), the United Nations supported the conservation and a wise use of wetlands worldwide.

Acronyms And Abbreviations

BRICS	Brasil, Russia, India, China, South Africa
CCS	Carbon Capture Sequestration
CHP	Combined Heat and Power
CFCS	Halocarbons
CSIRO	Commonwealth Scientific and Research Organization (Australia)
DNA	Deoxyribonucleic Acid
EFA	Ecological Footprint Analysis
E/R	Energy(r)evolution
adv E/R	Advanced Energy Revolution
ENSO	El Niño Southern Oscillation
EREC	European Renewable Energy Council
EU	European Union
FAO	Food and Agriculture Organization (United Nations)
FDA	Food and Drug Administration (USA)
FSU	Former Soviet Union
GDP (PPP)	Gross Domestic Product (Purchasing Power Parity)
GHG	Greenhouse Gases
GOE	Great Oxygen Event
GDP	Gross Domestic Product
GNP	Gross National Product
IEA	International Energy Agency
IMF	International Monetary Fund
MA	Millennium Ecosystem Assessment
NBER	National Bureau of Economic Research (Cambridge,Ma)
NGO	Non-Governmental Organization
OECD	Organization for Economic Cooperation & Development
PETM	Paleocene-Eocene Thermal Maximum
REF	Reference Model
RETD	Renewable Energy Technologies Deployment
RNA	Ribonucleic Acid

RTM	Reference Transition Model
SDP	Sustainable Development Process
SSE	Steady-State Economy
SSNNP	Sustainable Social Net National Product
UN	United Nations
UNESCO	UN Educational, Scientific & Cultural Organization
WCED	World Commission on Environment and Development
WEO	World Energy Outlook
WWF	World Wide Fund for Nature

Units Of Measurements

ky = 1000 years; **Bya** = billion years ago;
km (kilometre) = 1,000 m; **ha** (hectare) = 10,000 m^2
toe = energy obtained by burning 1 tonne of crude oil; **Mtoe** = million tonnes of oil equivalent[162]
Mil t/a = million Tonnes per annum; **Gt** = giga tonnes
W = measure of electrical capacity;
kW = 1,000 Watts; **MW** = 1 million Watts;
GW = 1 billion Watt (1 Gigawatt = 109 Watts); **TW** = 1 Trillion Watt (1 Terawatt = 1012 Watts)
W/m^2 = Watt per square meter; **J** = Unit of measure of Energy
PJ/a (Peta Joule per annum) = 1015 Joules per annum;
BTU = British thermal unit
kWh is a unit of energy equal to 1,000 Watts per hour

---

162 1 Toe is the energy released by burning 1 tonne of crude oil. IEA/OECD define 1 toe = 41.868 GJ. Conversion factors are used to express in Toe the combustion of diesel, biodiesel and bioethanol. Source Tonne of oil equivalent (Wikipedia)